Govardhan Reddy Patancheru

A wearable prototype of reflective sensor for non invasive measurement of heart rate

Anchor Academic
Publishing

Patancheru, Govardhan Reddy: A wearable prototype of reflective sensor for non invasive measurement of heart rate, Hamburg, Anchor Academic Publishing 2014

Buch·ISBN: 978·3·95489·335·5
PDF·eBook·ISBN: 978·3·95489·835·0
Druck/Herstellung: Anchor Academic Publishing, Hamburg, 2014

Bibliografische Information der Deutschen Nationalbibliothek:
Die Deutsche Nationalbibliothek verzeichnet diese Publikation in der Deutschen Nationalbibliografie; detaillierte bibliografische Daten sind im Internet über http://dnb.d·nb.de abrufbar

Bibliographical Information of the German National Library:
The German National Library lists this publication in the German National Bibliography. Detailed bibliographic data can be found at: http://dnb.d·nb.de

© Anchor Academic Publishing, ein Imprint der Diplomica® Verlag GmbH
http://www.diplom.de, Hamburg 2014
Printed in Germany

Abstract

Despite having the numerous evolved heart rate measuring devices and progress in their development over the years, there always remain the challenges of modern signal processing implementation by a comparatively small size wearable device. This research paper presents a wearable reflectance photoplethysmography (PPG) sensor system for measuring the heart rate of a user both in steady and moving states. The size and, power consumption of the device are considered while developing, to ensure an easy deployment of the unit at the measuring site and the ability to power the entire unit with a battery .The selection of both the electronic circuits and signal processing techniques is based on their sensitivity to PPG signals, robustness against noise inducing artifacts and miniaturization of the entire measuring unit. The entire signal chain operates in the discrete-time, which allows the entire signal processing to be implemented in firmware on an embedded microprocessor. The PPG sensor system is implemented on a single PCB that consumes around 7.5mW of power. Benchmarking tests with standard heart rate measuring devices reveal that the developed measurement unit (combination of the PPG sensor system, and inertial measurement unit (IMU) developed in-house at Acreo Swedish ICT, and a battery) is comparable to the devices in detecting heart rate even in motion artifacts environment.

This research work is carried out in Acreo Swedish ICT, Gothenburg, Sweden in collaboration with MidSweden University, Sundsvall, and Department of Electronics Design. This report can be used as ground work for future development of wearable heart rate measuring units at Acreo Swedish ICT.

Keywords: PCB, PC, photoplethysmography, motion artifacts, PPG signals, benchmarking.

Acknowledgements / Foreword

The largest lifetime is like a multi electromagnetic wave. However, only at some point large amplitudes (peaks) occur. I feel this research at Acreo Swedish ICT seats like a peak in my life. I would like to express my gratitude, appreciation and humble THANKS to everyone at the Acreo Swedish ICT sensor systems department, who contributed to this project's success directly or indirectly.

Firstly, I would like to THANK Peter Björkholm, the Acreo Swedish ICT sensor system manager, for making me feel completely ease with his warm, welcoming nature and for the excellent resources provided for the research at Acreo.

THANKS for my supervisor, Erik Svensson, at Acreo, for considering me for this research. His advice on research work and career has been priceless. His belief in me was by far, the most invaluable gesture that I have ever come across in my academic and professional life.

A big THANKS to my assistant supervisor, John Rösevall, at Acreo, for sharing his innovative ideas and enormous amounts of patience in considering seriously whatever proposals or approaches I randomly suggested throughout the journey.

I would like to THANK my supervisor, Johan Sidén, at Mid Sweden University for his valuable support in monitoring and encouragement throughout the research work.

And finally to my family members that encouraged me throughout the research.

Table of Contents

List of Figures

List of Tables

Notation

ADC	Analog-to-Digital Converter
AFE	Analog Front End
AFE-PPG	Analog Front End- Photoplethysmogram
ANC	Adaptive Noise Cancellation
BPM	Beats per Minute
CCS	Code Composer Studio
DST	Discrete Saturation Transform
GUI	Graphical User Interface
IAR	Ingenjörsfirman Anders Rundgren
ICA	Independent Component Analysis
IMU	Inertial Measurement Unit
PCB	Printed Circuit Board
PC	Personal Computer
PPG	Photoplethysmogram
RMSE	Root Mean Square Error
SOC	System on Chip
TI	Texas Instruments
USB	Universal Serial Bus
USCIs	Universal Serial Communication Interfaces

1 Introduction

Health monitoring has been an important field of interest over decades. The rise of need to continuous measure and assess all standard vital signs remotely and to monitor their trend over time is essential for the development of physiological tele-monitoring [1]. The vital signs being the heart rate, respiratory rate, blood pressure, body temperature, and oxygen saturation level in blood (SPO2). The emerging advances in the field of electronics, particularly with hardware miniaturization of devices measuring vital signs led to the development of wearable devices[Figure 1].For example, the development of compact and light-weight wearable devices could facilitate remote noninvasive monitoring of vital signs. The main benefits of deploying the mobile technologies in the field of medical care are [2]:

- Improve patient safety
- Decrease the risk of medical errors
- Increase physician productivity and efficiency

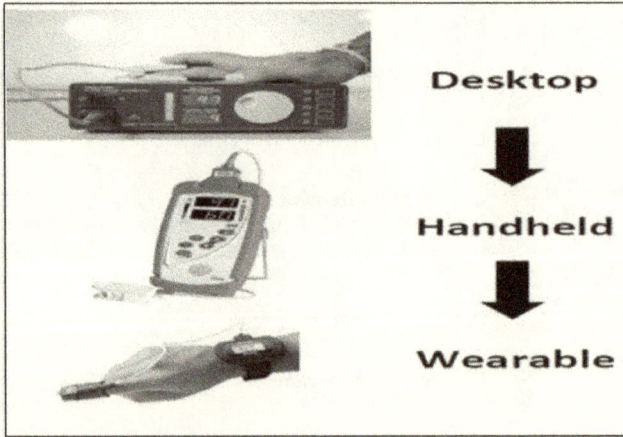

Figure 1: Reduction in the size of pulse meters by the advancement in the development [1]

1.1 Background and problem motivation

There have always been advances and improvements to the developed things in every field with the growing technology. When it comes to the field of health monitoring, the advancements in technology has led to the development of different types of heart rate measuring devices. The available heart rate measuring devices can be categorized into the following two types:

1. Measures heart rate only when the user is steady, any kind of user's movement will result in the erroneous heart rate measurements.
2. Measures heart rate of the user during in steady and moving states by using special techniques to detect motion artifacts and to reject them. The techniques used by these devices have high computational complexity. They also exhibit issues with the sensor attachment.

There is a real need to develop a wearable heart rate measuring unit to overcome the limitations of available heart rate measuring devices.

1.2 Overall aim

The aim of the research work is to develop a miniaturized wearable PPG sensor system for measuring PPG signals and a low computational complexity algorithm to measure the heart rate of a user. And to develop a MATLAB graphical user interface (GUI) to display the measured PPG signals and heart rate, when:

1. The measurements are carried out by connecting the measurement unit to a personal computer (PC) via Universal Serial Bus (USB) interface.

2. The measurements are stored on flash memory of IMU, which will be later read back into a PC using the USB serial communication between the measurement unit and PC.

1.3 Scope

The tasks needed for this research implementation are listed below:

1. Design and developing a sensor unit for measuring the PPG signals.

2. Sampling PPG signals of sensor unit by on-board microcontroller of IMU developed in-house at Acreo Swedish ICT.

3. Processing the sampled signals on IMU to calculate the heart rate.

4. Establishing a USB serial communication between the measurement unit and a PC.

5. Developing a MATLAB GUI to display the measured PPG signals and heart rate.

1.4 Concrete and verifiable goals

Study different heart rate measuring devices to build a low power consuming and miniaturized PPG sensor system.

Study the effect of motion artifacts on the measurements and to implement an algorithm for removing the motion artifacts. Verifying the device working with other standard heart rate measuring devices

1.5 Outline

The following presents the summary of each of the remaining chapters:
   1. Theory / Related work gives the knowledge needed for a reader.

2. Methodology illustrates the methods used for the development.

3. Implementation and Results presents the implementation process of both hardware and software along with their results. It also presents the achieved results comparison with the other pulse meter devices.

4. Conclusions and Future work provides an executive summary of the main achievements of this project. It also discusses the future work.

1.6 Contributions

The investigations, design of PPG sensor system and mounting the components on the designed PCB board was carried out by the author. The PCB of the PPG sensor system was designed by a hardware engineer of Acreo Swedish ICT. The techniques for removing motion artifacts were investigated and implemented by author. The program used for saving the data on flash memory was developed by Acreo Swedish ICT.

Integration of PPG sensor system with the IMU board, serial communication between the IMU and host PC for providing a user interface for monitoring the real time heart rate was also carried out by the author.

2   Theory / Related work

This chapter discusses the background and introduces the reader to the related work.

2.1 Introduction to photoplethysmography

Photoplethysmography refers to the non-invasive measurement of blood volume in a specified region. The volume of blood in a specified region increases in the systole phase and decreases in the diastole phase during the cardiac cycle of heart as illustrated in Figure2.1. This changing blood volume can be directly used to calculate the heart rate and also to measure other characteristics of cardiovascular function.

The basic PPG sensing system consists of a light source to illuminate the blood vessels and a photo detector to sense the received light that is a result of optical absorption and scattering properties of the blood, tissue and bone. The PPG signal consists of two components referred to as AC and DC as shown in Figure2. The AC component is caused by the pulsatile changes in arterial blood volume and is synchronous with the heart beat because of which it can be used as a source for the heart rate information. The DC component is caused by the tissues and average blood volume that superimposes with the AC component. The DC component should be removed from the whole signal to get desired information of heartrate from the AC component.

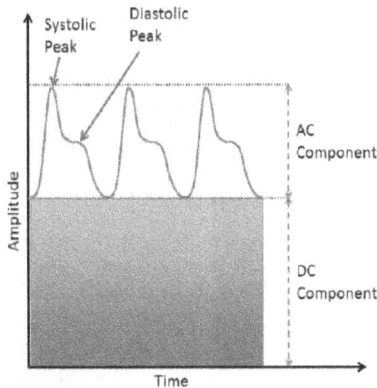

Figure 2: Changes in blood pressure [30]

2.2 Construction of photoplethysmography sensor system

Construction of wearable PPG sensor system depends mainly on the following two factors:
The location of the sensor and the way it is attached to the user.

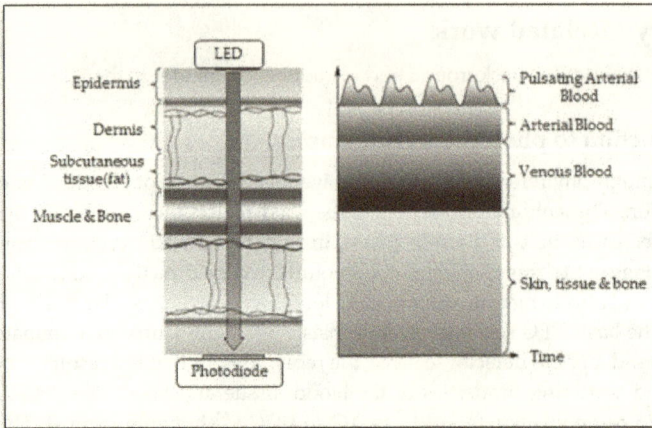

Figure 3: Transmission mode of Photoplethysmography [1]

A PPG sensor can be placed at any place that has a blood flow. Depending on the location of the sensor, the construction of PPG sensor system can be made in the following two different modes:

Transmission mode: The photo detector and LED are placed on the opposite sides of the tissue to be measured. The photo detector measures the amount of light that was not absorbed as illustrated in the Figure3.

Reflection mode: The photo detector and LED are placed on the same side of the tissue to be measured. These measures the amount of light backscattered from the skin and capillaries. This is illustrated in the below Figure4.

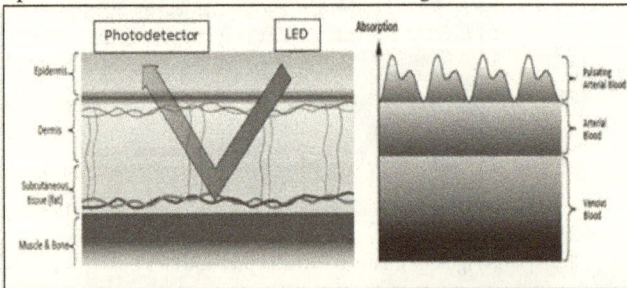

Figure 4: Reflection mode of Photoplethysmography [1]

2.2.1 TRCT1000-phototransistor with infrared LED

The TRCT1000 is a reflective optical sensor that has included both the infrared light emitter of wavelength of 950nmand phototransistor side by side in a leaded package such that it has less effect from the surrounding visible light.

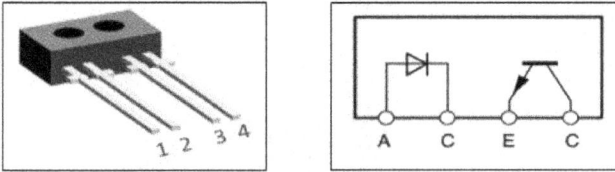

Figure 5: TRCT1000 photo transistor

2.2.2 NJL5303R-phototransistor with green LED

The NJL5303R is a reflective optical sensor that includes both the green LED of wavelength of 570nm and a photo transistor in a small package that are well suited for pulse detection. In general, the green light has a higher reflective factor than the factor of infrared light which provides more sensitive detection and high signal to noise ratio [10].

The NJL5303R's green phototransistor has high sensitivity to the measuring pulse waves as illustrated in Figure6; with a green circle corresponding to the wavelength of the green wavelength spectrum.Figure7 depicts the working principle of NJL5503R.

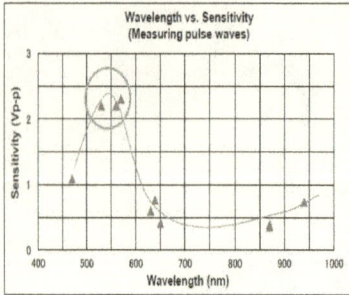

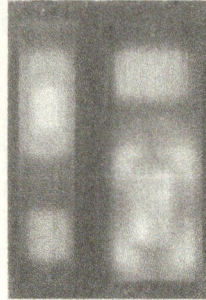

(a) Sensitivity of photo reflector   (b) Pictorial view of the photo reflector

Figure 6: NJL5503R photo reflector [10]

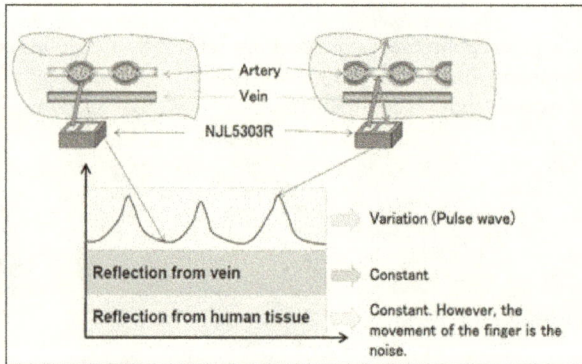

Figure 7: Working principle of NJL5503R [10]

2.3 Issues with PPG signal measurements

In general, measurements of PPG signals will be affected by different factors as listed below:

2.3.1 Artifacts

In general, artifact refers to the disturbance in the measured PPG signal. The two types of artifacts associated with the PPG signal measurements are explained below:

2.3.1.1 Ambient artifacts

The light sources other than the LED included within the PPG sensor system results in the ambient light artifacts. The indoors fluorescent/incandescent lighting forms the source of ambient light artifacts if the measurement is done in the laboratory environment. The other main source will be the sun's light, either coming through a window or from the sensor being worn while the user is outdoors [1].

The sources of artificial light will be generated from the electrical mains supply having a fundamental frequency of 50Hz or 60 Hz.

2.3.1.2 Motion artifacts

Motion artifact is any corruption of the PPG signal due to the user's motion [1]. Motion artifacts will be resulted from the mechanical distortion of the optical path between the LED and photodiode of the PPG transducer [1]. This type of mechanical distortion comes into picture when the measurements are carried out by placing the PPG sensor on forehead, during which there exists changes in the relative position of the sensor with respect to the frontal bone of the skull rather than relative movements of the sensor with respect to the skin. This mechanism results in the changes of distribution of LEDs backscattered light reaching the photo sensor, thus leading to the corruption of the PPG signal.

2.3.2 Pressure disturbances acting on the PPG sensor

A too low contact pressure between the PPG sensor and measurement site will result in distorted PPG signals leading to inaccurate measurements [7]. On the other hand, a too high contact pressure may result in comprise of blood circulation when the measurements are conducted for a longer time thus leading to the complete loss of PPG data.

2.3.3 Physical activity of the user

Its sources can be 1) the formation of air gaps created between the skin and sensor during the physical activity of user, 2) variation in venous pressure resulted from the back and forth movement of a user's physical activity[2].

2.4 Minimizing the problems associated with PPG measurements

Following describes the techniques to minimize the effect of problems associated with PPG signal measurements:

2.4.1 Minimizing motion artifacts

According to different studies conducted to overcome the effects of movement artifacts, suggested different methods to improve the measurement accuracy when the user is steady while leaving out the limitations with the measurements during motion artifacts. The following explains the ways to minimize the motion artifacts:

2.4.1.1 Measurement site

The artifacts explained in the section 2.3.1.2 are dependent on the measurement site. The study made by Mendelson [3] states that the reflected sensor located on the forehead provides more consistent results when the user is motionless (steady) and when a moderate amount of motion artifacts are present as compared to the measurement carried out in other facial regions. The study of Mannheimer [4] reveals that the placement of the sensor directly over the eyebrow slightly lateral to the iris also provides consistent measurement results.

2.4.1.2 Signal processing

Signal processing is the most common used methods to overcome the problems of motion artifacts. There have been numerous algorithms developed for motion artifact removal, including Independent Component Analysis (ICA) and Adaptive Noise Cancellation (ANC) that can be applied in general to any sensing systems to remove noise. The specific methods for solving the motion artifacts of PPG sensor unit as explained by J.A.C.Patterson [1] are

1) Discrete Saturation Transform (DST) used by the Masimo which is a leading commercial pulse oximeter manufacturer,

2) Wavelength method proposed by Hayes and Smith,

3) Wavelet transforms method proposed by Addison and Watson.

All the above mentioned techniques are very computationally intensive solutions for detecting and removing the motion artifacts. There is a need to develop a low computational complexity technique for detecting and minimizing the motion artifacts.

2.4.1.3 Sensor attachment

Attachment of sensor on the proper measurement site also plays a major role in minimizing the effect of motion artifacts. The three widely used methods are [Figure 8]; the first method uses an adhesive tape for attachment and the second uses a headband for the sensor attachment. The third method is to embed the sensor into pre-existing equipment's like a soldier's helmet or goggles. According to various researches, the usage of compressive headband will be the optimal choice as it presents low pressure venous pulsations and venous pooling when the user is in Trendelenberg position [2], where the user's body will be lying flat on the back with the feet higher than the head by 15-30 degrees.

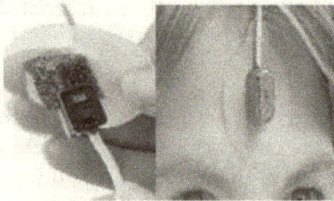

(a)Adhesive attachment [10]   (b) Headband attachment [8]

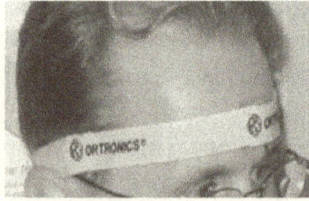

(c)Helmet Integrated sensor[9]
Figure 8: Methods of sensor attachments

Even though each method has its own advantage over others, they also exhibit some degree of motion artifacts. When it comes to the adhesive tape method, the perspiration effect on adhesive tape will possibly make the sensor to lose its contact with skin thus resulting in increased motion artifacts. In case of headband method, there is a possibility of slipping the sensor and headband from the actual measurement site resulting in the inability to measure the valid PPG signals.

2.4.1.4 Sensor contact pressure

There were several studies conducted to investigate the effect of contact pressure on the PPG signals and the studies here [5], [6] provide a qualitative description of the optimal contact pressure needed for the valid PPG measurements.

2.5 Inertial Measurement Unit

In general, the term IMU refers to an electronic unit that consists of accelerometers and gyroscopes for measuring on a craft's velocity, orientation and gravitational forces. It forms the main component of inertial navigation system mainly used in aircrafts, spacecraft's and watercrafts.

For this research implementation an IMU developed in-house at Acreo Swedish ICT as shown in Figure 9 was used. It has been integrated with various modules used for other specific applications.

Figure 9: IMU developed at Acreo Swedish ICT

2.5.1 Microcontroller

The CC430 provided by Texas Instruments (TI) is an ultra-low-power microcontroller system-on-chip (SOC) with integrated RF transceiver cores. Figure 10 shows the pin configuration of CC430F513x.

The one used here from such a family of CC430 is CC430F5137 microcontroller. It is a microcontroller SOC configuration that combines the following

- Sub-1-GHz RF transceiver with the MSP430 CPUXV2
- 32KB of in-system programmable flash memory up to 4KB of RAM,
- Two 16-bit timers, a high performance 12-bit ADC with six external inputs plus internal temperature and battery sensors,
- A comparator, Universal Serial Communication Interfaces (USCIs),
- A 128-bit AES security accelerator,
- A hardware multiplier,
- A DMA,
- An RTC module with alarm capabilities, and up to 30 I/O pins.

RGZ PACKAGE
(TOP VIEW)

Top pins (left to right):
P2.3/PM_TA1CCR2A/CB3/A3
P2.4/PM_RTCCLK/CB4/A4/VREF-/VeREF-
P2.5/PM_SVMOUT/CB5/A5/VREF+/VeREF+
AVCC
P5.0/XIN
P5.1/XOUT
AVSS
DVCC
RST/NMI/SBWTDIO
TEST/SBWTCK
PJ.3/TCK
PJ.2/TMS

Left pins:
P2.2/PM_TA1CCR1A/CB2/A2 — 1
P2.1/PM_TA1CCR0A/CB1/A1 — 2
P2.0/PM_CBOUT1/PM_TA1CLK/CB0/A0 — 3
P1.7/PM_UCA0CLK/PM_UCB0STE — 4
P1.6/PM_UCA0TXD/PM_UCA0SIMO — 5
P1.5/PM_UCA0RXD/PM_UCA0SOMI — 6
VCORE — 7
DVCC — 8
P1.4/PM_UCB0CLK/PM_UCA0STE — 9
P1.3/PM_UCB0SIMO/PM_UCB0SDA — 10
P1.2/PM_UCB0SOMI/PM_UCB0SCL — 11
P1.1/PM_RFGDO2 — 12

CC430F513x

Right pins:
36 — PJ.1/TDI/TCLK
35 — PJ.0/TDO
34 — GUARD
33 — R_BIAS
32 — AVCC_RF
31 — AVCC_RF
30 — RF_N
29 — RF_P
28 — AVCC_RF
27 — AVCC_RF
26 — RF_XOUT
25 — RF_XIN

Bottom pins (left to right):
P1.0/PM_RFGDO0
P3.7/PM_SMCLK
P3.6/PM_RFGDO1
P3.5/PM_TA0CCR4A
P3.4/PM_TA0CCR3A
P3.3/PM_TA0CCR2A
P3.2/PM_TA0CCR1A
P3.1/PM_TA0CCR0A
P3.0/PM_CBOUT0/PM_TA0CLK
DVCC
P2.7/PM_ADC12CLK/PM_DMAE0
P2.6/PM_ACLK

VSS
Exposed die
attached pad

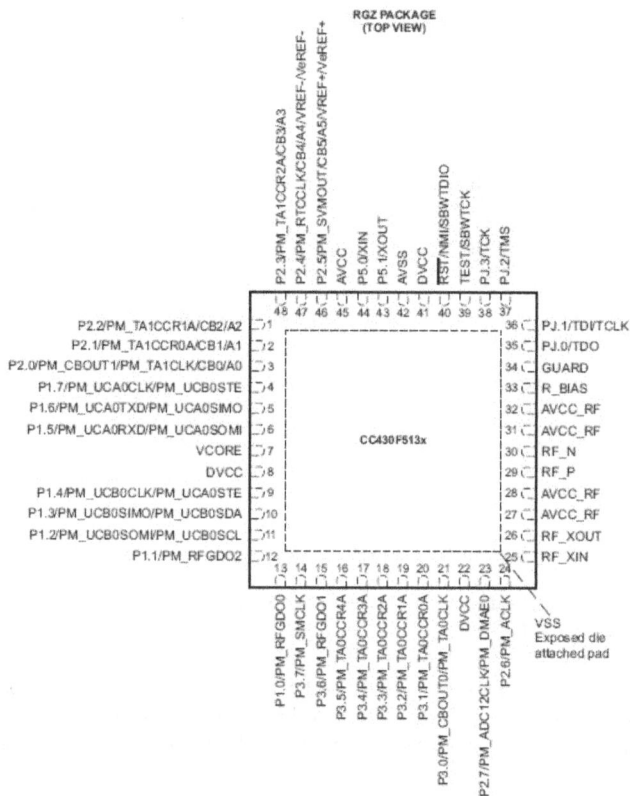

Figure 10: Pin configuration of CC430F513x [28]

2.5.2 Compiler and debugger

The code composer Studio (CCS) developed by the TI is the compiler supported by all the TI embedded processor families. It is used as the compiler for the CC430F5137 microcontroller.

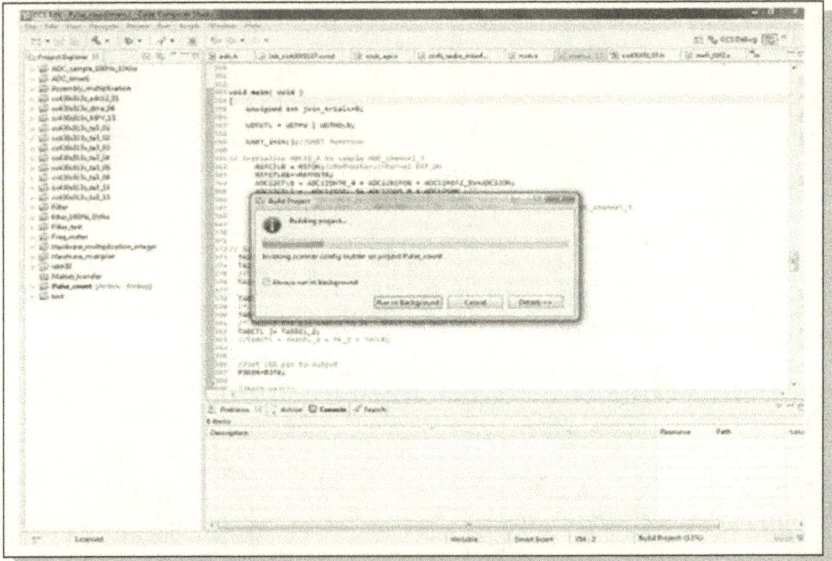

Figure 11: A still of the CCS window

The TI's MSP-eZ430 USB stick [Figure 12] is a complete development tool providing all the software and hardware to evaluate the CC430f5137.It supports both Ingenjörs-firman Anders Rundgren (IAR) embedded workbench and CCS environment. It runs on the power supplied by the USB port thereby eliminating the need of external power supply.

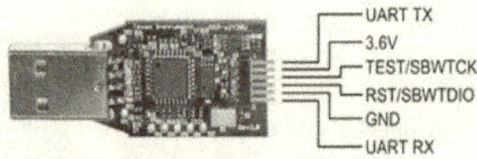

Figure 12: MSP-ez430 Debug Interface

2.6 Heart rate monitoring in typical exercises and sports

Monitoring heart rate in typical exercises like running, walking, jogging and cycling is a traditional way to improve the performance of the user. For example, walkers can use the measured heart rate for adjusting the intensity of their walk thus allowing them to speed up or slow down to stay in their chosen heart rate zones.

In sports, monitoring heart rate of the athlete helps the trainer to understand the needed prescriptions of the particular intensity and duration blends to achieve a successful performance of the athlete.

2.6.1 Heart rate monitoring devices for typical exercises

The different kinds of heart rate monitors (HRM) available for monitoring the heart rate of a user during typical exercises are listed below.

2.6.1.1 Heart rate monitors with a chest strap

Most of the available heart rate monitors use a chest strap that fits snugly around the chest. The detected electrical activity of the heart will be transmitted to the display, usually worn like a wrist watch or transmitted to the mobile apps via Bluetooth. Working of these devices is limited with its attachment to the measuring site. They can result in the heart rate reading lying outside the normal heart rate range of 40-200beats when the strap is not connected properly to the measurement site. The low-end chest-strap models can have interference with the other HRM's [25]. An example of this device is shown in Figure 13.

Figure 13: RF Polar H6-Heart rate monitoring device [19]

2.6.1.2 Heart rate monitors without a chest strap

These consist only of a wristwatch-style monitor that uses optical sensors on the back of the device to continuously read the pulse through the skin providing real time heart rate. An example of this device is shown in the Figure 14.

Figure 14: Polar FT7-Heart rate monitoring device [19]

2.7 Measuring heart rate of a swimmer

Like the other typical exercise activities (running, cycling, jogging, etc.), swimming also burns lots of calories, supports weight, builds muscular strength and improves the cardiovascular fitness. It is recommended for people with joint problems or with overweight as it takes a great deal of strain off of the skeleton. HRM's are being used to monitor the heart rate of a swimmer.

2.7.1 Methods to improve the swimming performance

Performance of swimmer can be improved by following interval-based swim sets instead of continuous swim sets. For example, instead of swimming a straight 1000-mter freestyle (40 lengths in a 25 meter pool), swimming should be done in a set of 20 x 50's freestyle (two lengths of the pool freestyle, 20 times) on a particular timed interval [20].

Interval sets are designed in such a way that it allows the swimmer to have rest time and recover after each individual swim within a particular set which in turn has the following benefits with the added rest and recover[20]:

1. It helps in building endurance.

2. It gives the room for the swimmer to maintain proper stroke technique by remaining fresh throughout the set.

3. It also allows the swimmer to challenge them by increasing the effort during a set of without becoming too fatigued.

The interval based set can be designed with the following two methods.

1. Interval sets with a specific amount of time or rest between swims

An example for this swimming method would be 10x50's freestyle (meters) with 15 seconds rest between each 50 in a way the swimmer is swimming a 50-yard freestyle 10 times resting 15 seconds between each swim. This method can be best method to focus the heart rate training as it always leaves the same time interval during rest for monitoring the heart rate of swimmer.

2. Interval sets with a specific time to leave for each swim

This method sets a specific interval or time for each set that includes both the swimming and the rest activity. An example for this method would be a set of 10x50's freestyle (meters) such that each set of 50-meter swim and rest will be finished in 60 seconds. In detail, it can be said that if a swim was completed in 45 seconds then the remaining 25 seconds can be the rest time.

2.7.2 Devices for monitoring heart rate of a swimmer

The following lists out the challenges for constructing the devices for monitoring the heart rate of a swimmer when compared to the construction of heart rate monitoring devices for typical exercise of running, walking and jogging.

1. The device should be water resistance

2. Small in size such that the swimmer is willing to wear

3. Sensor should not be prone to hydrostatic (water) pressure because it might result in the corrupted PPG signal measurements.

There are different devices available for monitoring the heart rate of swimmer during swimming [23] and some of them are explained below.

2.7.2.1 Suunto memory belt

Suunto memory belt is a chest belt that records the heart rate and stores it on the integrated memory chip [Figure 15]. The data stored on the memory chip can be downloaded and analyzed with the Suunto PC software to get the information the heart rate variation [22].

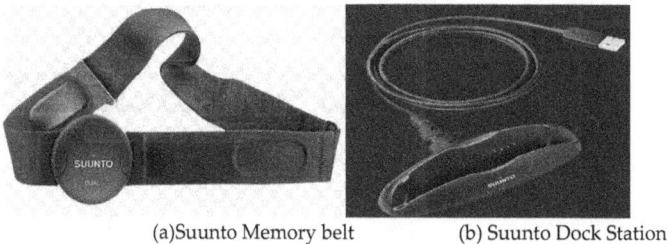

(a)Suunto Memory belt   (b) Suunto Dock Station

Figure 15: Suunto heart rate monitoring device [21].

Drawback:

1. There might be chances of strap sliding down to the athlete's waist after speedy push-off resulting in the wild readings.

2. Sometimes wearing of chest belt can be felt unnatural and can hinder movements.

2.7.2.2 Suunto ANT heart rate belt

Operating on a 2.4 GHz frequency, the Suunto ANT device [Figure 16] transmits the heart rate information from the chest strap to the other Suunto device being used by coach or trainer [25].

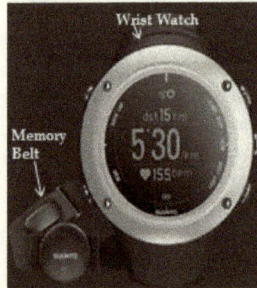

Figure 16: Suunto ANT heart rate monitoring device [25]

Drawback: The device doesn't get connected again, if a break in the wireless transmission occurs which in turn does not show the real time heart rate information at the coach, trainer or researcher end.

2.7.2.3 Instabeat

Instabeat [Figure 17] is a pretty simple device attaching to the swimming googles that measures and displays the heart rate, calories burnt, laps and turns during the swim on the goggles.

The best part is that it automatically turns on optical sensor when it is placed on the head and projects a color onto the lenses of the user. Each color has three levels that indicate the user's beginning, middle and upper limit of a particular zone (i.e., the heart rate, calories burnt, laps or turns).

(a)Instabeat device [23]   (b) Swimmer with Instabeat goggles [24]

Figure 17: Instabeat-Heart rate monitoring device

Drawback: It is specially designed for the swimming application and cannot be used at the other measurement sites like finger or wrist.

3   Methodology

This chapter explains the various methods employed for developing the heart rate measurement unit.

3.1   Measuring PPG signals

From chapter 2, it can be concluded that the PPG signal will be affected by different factors, so the system should be designed such that it minimizes these factors to get a valid measurement.

3.1.1 Proposed methods for building reflective PPG system

Apart from the things explained in section 2.2.1, distance between the LED and photo sensor also plays a vital role in the output of the PPG sensor system. To overcome the issue of experimenting methods to find out the best distance, the choice made here was to use a component that has built-in phototransistor and LED with a fixed distance between the two, rather than using a separate LED and a phototransistor. The following are the two such components as explained in section 2.4 and will be used in the investigation to find a better reflective PPG system for the application requirements.

1) TCRT1000-Phototransistor and infrared LED

2) NJL5303R-Phototransistor and green LED

3.2   Improving the raw PPG signal for desired Heart Rate

The output of photo transistor will be the raw pulse signal that is a combination of the desired pulse signal with the undesired DC and AC components as explained in chapter 2.These undesired components should be ruled out or minimized to make the desired pulse signal.

3.2.1 Minimizing the DC and AC component

The DC component can be removed by using the high pass filter, while the unwanted AC components as explained in the section 2.3.1.1 can be minimized by using a suitable low pass filter such that it allows the frequencies needed for the desired heart rate.

3.2.2 Frequency limits for the desired heart rate

The heart rate typically expressed as Beats Per Minute (BPM) can be calculated by counting the frequencies measured from the time interval between the two pulse beats over the 60 second interval.

$$\text{HeartRate(BPM)} = f_{measured} * 60$$

Desired heart rate to be measured lies in the range of 42 BPM to 144 BPM that correlates to a frequency range of 0.7Hz to 2.4Hz.Thus the required application needs a bandpass filter that allows the frequencies in the range of 0.7Hz to 2.4Hz.

3.3 Sampling the AFE- PPG signal by the microcontroller

The foremost thing of sampling raw PPG signal by the microcontroller is needed before processing the raw PPG signals for calculating the heartbeat rate. Figure 18 illustrates the flowchart to sample an analog signal.

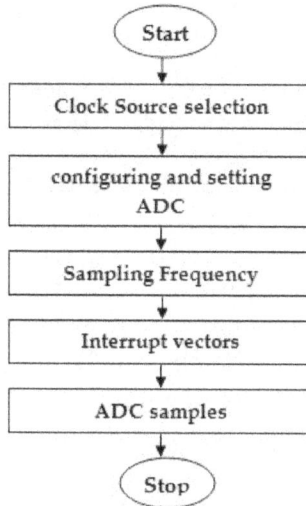

Figure 18: Method to read AFE-PPG signal by the microcontroller

3.4 Digitizing the sampled PPG signals

Each ADC sampled value of the analog signal will be representing the digital equivalent value of the analog signal. The digital value of ADC, which depends on the selected bit resolution and ADC reference voltage, will be converted to the respective voltage of the analog signal.

Finally, the digital signal of the microcontroller signal will converted to the digital pulse signal as illustrated in Figure 19 for the post processing of heart rate.

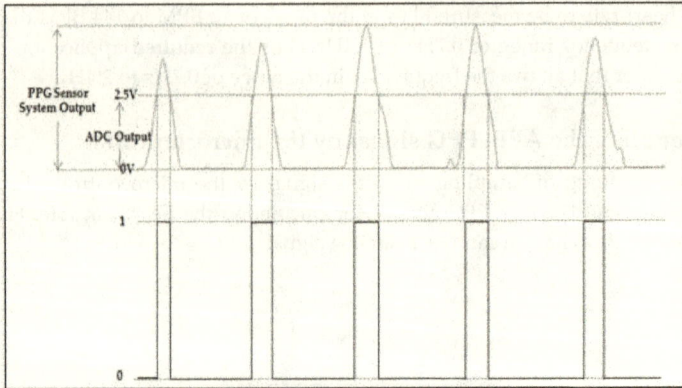

Figure 19: Conversion of sampled PPG signals to digital signal

3.5   Processing digital signals to calculate heart rate

As explained in section 2.2.1, the heart rate is the number of heart beats counted in a minute and this can be worked out by quantifying the frequency of the PPG signals. Methods for determining the frequency of a signal are listed below and are illustrated in the Figure 20.

1. Considering the number of HIGHs and LOWs in one minute.

2. Considering the time between the two consecutive HIGH-to-HIGH transitions

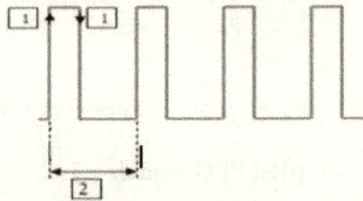

Figure 20: Methods to determine the frequency

3.5.1 Technique for minimizing motion artifacts

There is a need to remove the frequency components, that falls out of the 0.7 to 2.4Hz frequency range, corresponding to the range of heart rate of 42BPM to 144BPM.As explained in the section 2.4.1.2, there exists a number of techniques for dealing the artifacts but the one employed for this application is to identify and reject corrupted PPG signals based on the time interval between the two consecutive pulses.

3.6 Power supply

The prototype to be developed should be supplied with power both when the measurements are taken with the user activities like resting, sitting, standing and with the activities of user involved with movements like running, jogging, cycling, etc. To meet the requirements, the proposed design includes a micro-USB connection that provides the necessary power when plugged to a PC. On the other hand, for taking the measurements of a user with activities involved movements, a lithium-ion battery is employed to accomplish the task of taking measurements for a long time such as some hours or even a day for saving information on storage module. The micro-USB also facilitates to recharge the battery when required. According to [11], the lithium-ion battery delivers the highest power density of all batteries available on the commercial market on a per-unit-of-volume basis.

3.7 Measurement unit reliability

The various factors affecting the PPG signals are listed out in section 2.3 and among those the factor that affects the PPG signals on a larger ratio is the movement of the user. As explained in the section 2.3.1 the PPG sensor measurements will be affected by motion artifacts resulting in the measurement "gaps". There might be chance of providing inaccurate results if these interval durations become too long. Thus, for proving the feasibility of measurement unit there is a need to show that the long gaps occur with a small probability such that the heart rate to be monitored from the measured PPG signal won't be affected on a large scale. Figure 21 illustrates the measurement gaps.

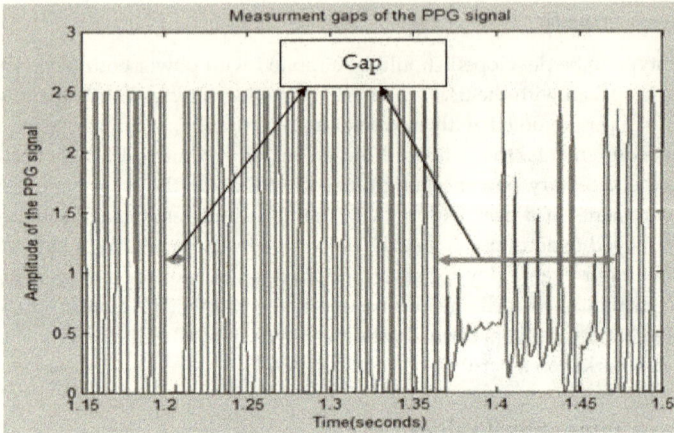

Figure 21: Measurement gaps formed by the biking activity of user

3.8 Probabilistic model of PPG sensor performance

To deal the challenges associated with the measurement gaps and results, the measurement unit functioning will be transferred into the context of repairable systems [14].

A repairable system characterizes the state of a device as either functional or non-functional such that the period of time for which the sensor system will be functioning is termed as "uptime" and the time for which the sensor system is not functioning is termed as "repair time", where the device will be in the repair state until it encounters uptime again [14]. For the developed measurement unit, the uptime will be the time for which the PPG sensor is giving a proper reading and the time for which there exists a measurement gap that is resulted from the invalid PPG sensor reading is termed as repair time. Figure 22 illustrates the uptime and repair time of the measurement unit.

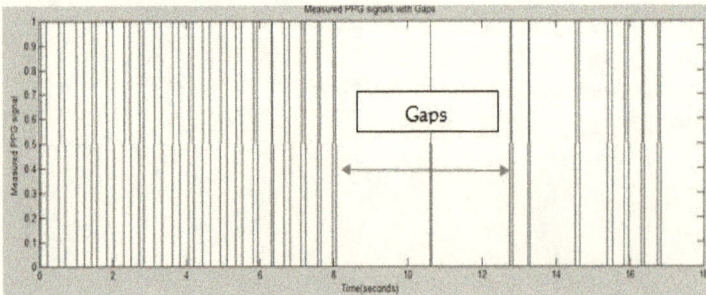

(a)Measurement gaps formed from the biking activity of the user

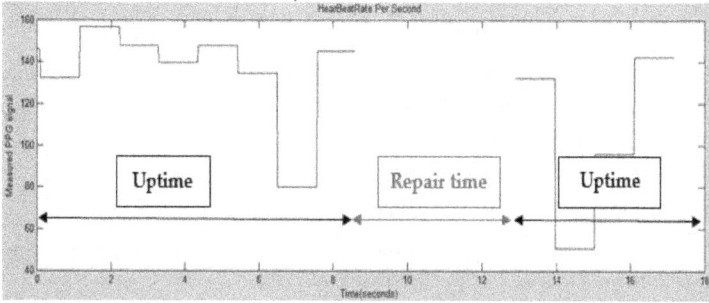

(b) Calculated heartrate

Figure 22: Uptime and repair time of the sensor system

4 Implementation and Results

This chapter explains the construction of the measurement unit along with the user study conducted to validate the prototype performance. It also explains the physiological and wearability requirements for using the unit at different measurement sites.

4.1 Block diagram of the prototype

Figure 23 shows the block diagram of the entire measurement unit connected to a PC, with each of them explained in the following sections

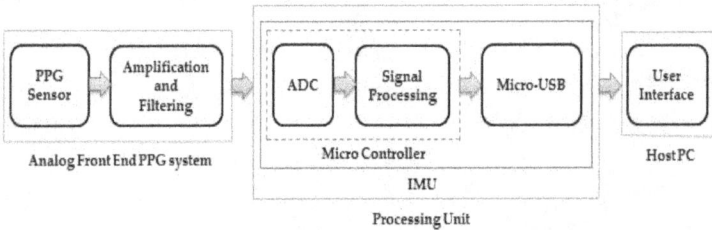

Figure 23: Block diagram of the measurement unit

4.2 Hardware implementation of the PPG sensor system

4.2.1 Design models of the PPG sensor system

The following presents the designing models of a PPG sensor system.

4.2.1.1 TCRT1000 Phototransistor-PPG sensor system

The circuit implementation of PPG system using TCRT1000 phototransistor is shown in Figure 24 (a). Using Rs=330 Ω and Rl= 10KΩ with a supply voltage of 3V, the TCRT1000 draws a current of 5mA and the resulted output is shown in Figure 24.

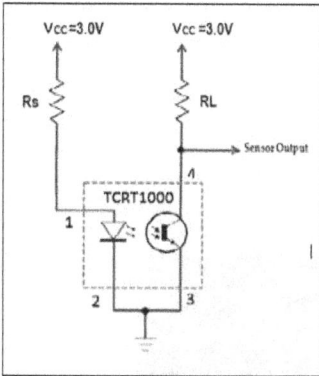

(a)Circuit of TRCT1000-PPG sensor system redrawn from [13]

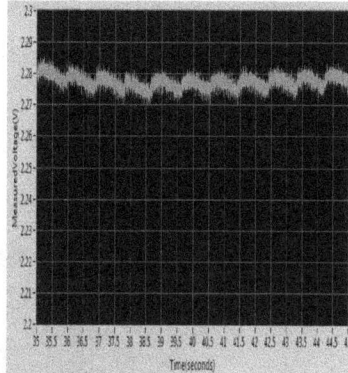

(b)Output PPG signal in labview

Figure 24: PPG sensor system implemented with TCRT1000

4.2.1.2 NJL5303R-PPG sensor system

The circuit implementation of PPG system using NJL5303R phototransistor is shown in Figure 25 (a). Using Rs=470 Ω and Rl= 5.5KΩ with a supply voltage of 3V the NJL5303R draws a current of 3mA and the resulted output is shown in the Figure 25.

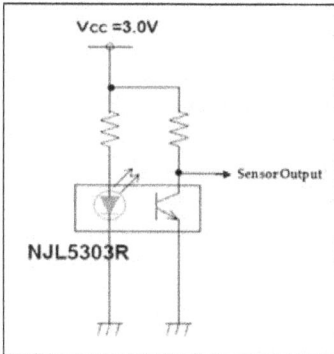

(a)Circuit of NJL5303R- PPG sensor System redrawn from [10]

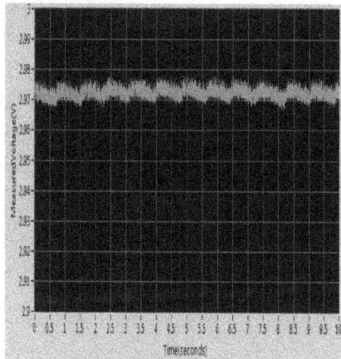

(b)Output PPG signal in labview

Figure 25: PPG sensor system implemented with NJL5303R

4.2.2 Improving PPG signal for the desired heart rate

The output of PPG sensor [Figure 26] has two components as explained in the section 2.1, with one the AC component that provides information about the heart rate and the other as the DC component of around 2.968V that is superimposed on the AC component. This clears the point that further processing of PPG signals is needed for filtering the DC component and to amplify the small portion of the AC component to extract the required information from it.

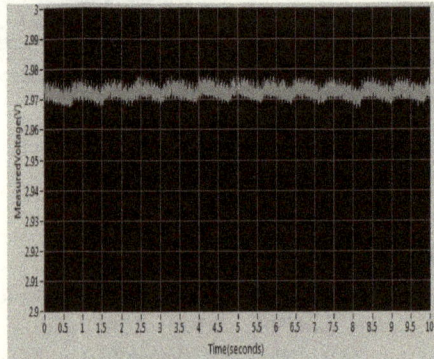

Figure 26: Output of NJL5303R phototransistor in lab view

4.2.2.1 NJL5303R circuit simulation in NI multisim

Figure 27 shows the circuit simulation in NI multisim, where the output of NJL5303R phototransistor collected from the section 4.2.1.2 was feeded as input to the high pass filter of first stage signal conditioning followed by the second stage signal conditioning [13].The simulated circuit is divided into three stages; the first stage of PPG sensor was explained in section 4.2.1 and the remaining two stages are explained in the following sections.

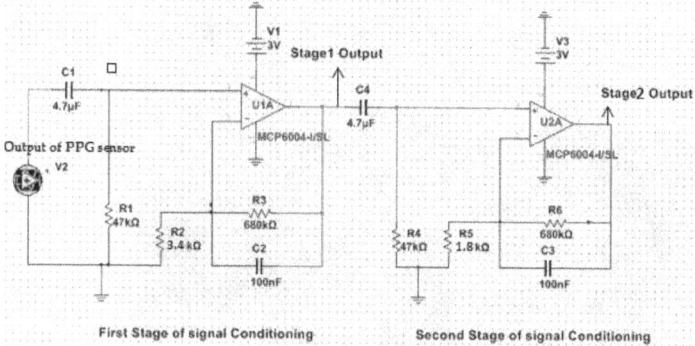

Figure 27: Circuit simulated in NI multisim

4.2.2.2 First stage of signal conditioning

The first stage of signal conditioning consists of a simple high pass filter followed by an active low pass filter as explained below.

High pass filter: It is an RC high pass filter used for filtering the DC component that has been superimposed with the AC component of the PPG signal. The cutoff frequency of 0.7Hz removes the frequencies lying below it, to achieve the desired lower limit of 42 BPM.

Active low pass filter: This filter stage has a cutoff frequency of 2.34Hz to remove the frequencies lying above it to achieve the upper limit of 140BPM of the desired heart rate and to remove the mains supply interference of 60 Hz or 50 Hz. The amplification factor of 200 amplifies the small portion of the AC component of the PPG signal as shown in Figure 28.

4.2.2.3 Second stage of signal conditioning

This stage has the similar high pass filter design as first stage and the low pass filter with a cutoff frequency of 2.34Hz with an amplification factor of 377.

Thus, the PPG signal resulting signal from the second stage will be the square pulses as shown in Figure 28.

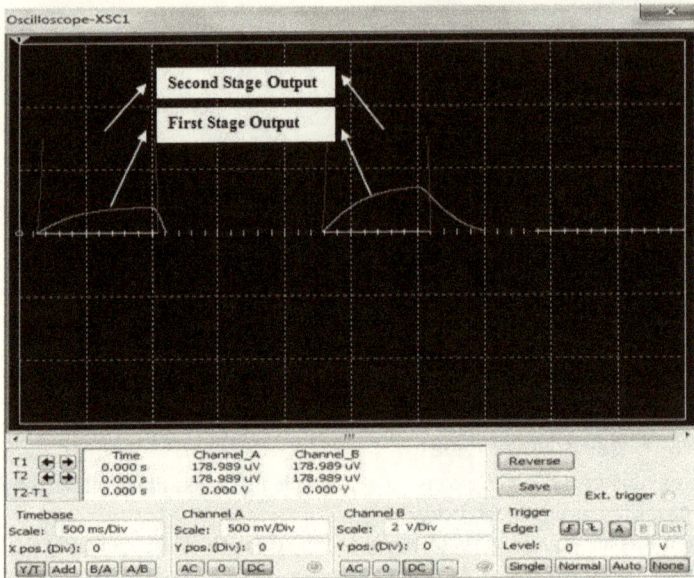

Figure 28: Simulation result in NI multisim

From Figure 28, it can be observed that the output of first stage is a pulse signal having maximum amplitude of 400mV and the amplification of second stage saturates the pulse signal at 3v that is the supply voltage of pomp resulting in square pulses.

4.2.3 Strip board implementation of the circuit

The selection of components for building the circuit simulated in NI multisim on strip board is explained in the following section with the implemented circuit shown in Figure 29.

4.2.3.1 Electronic component selection

PPG sensor: Even though both NRJ5303R and TRCT1000 output a similar PPG signal, they have differed in the prospects of power consumption and size. The TCRT100-PPG system produces a good output PPG signal at a minimum of 5mA and for the NJL5303R-PPG system it is 3mA. On the other hand, the TCRT100 component has a large dimensional size compared to the NJL5303R.The prototype to be developed is for wearable application, so this focuses more on miniaturizing the circuit; hence the NJL5303R has been finalized.

Voltage regulator: The Analog-to-Digital Controller (ADC) of CC430f5137 microcontroller can read a maximum voltage of 2.5V. Xc6215B302NR is selected as the voltage regulator for providing a supply voltage of 2.5 as it consumes power of 0.8uA.

Opamp: MCP6004-I/SL is selected as the opamp to use with the filtering stages as it consumes extremely a low current of 100 μA.

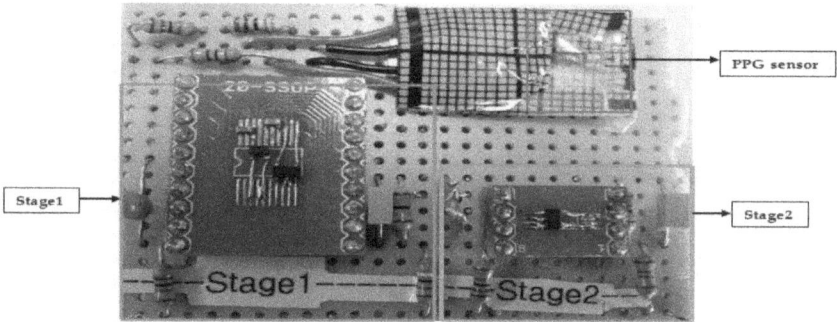

Figure 29: Strip board implementation of the circuit

The first and second stage output of the PPG sensor system is sampled using lab view and are as shown in the Figure 30.

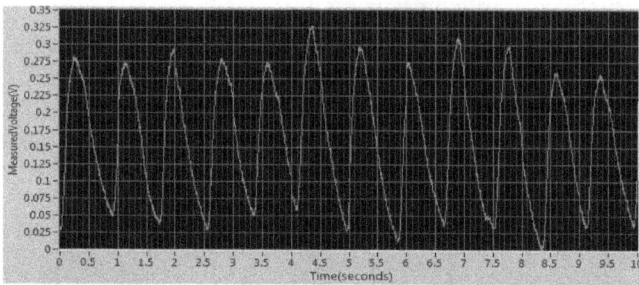

(a) Output PPG signal from the first stage signal conditioning

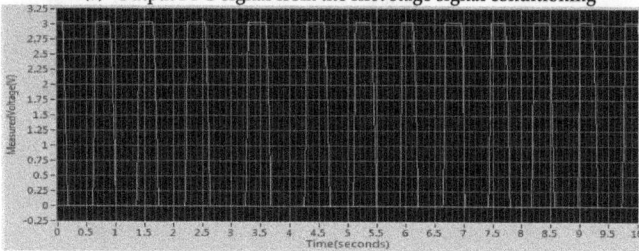

(b)Output PPG signal from the second stage signal conditioning

Figure 30: Outputs of the signal conditioning stages

Figure 30 represents output of first and second stage that was sampled at different time instants.

4.2.4 PCB of the Analog Front End-PPG sensor system

Figure 31shows the implemented PCB of the Analog Front End (AFE)-PPG sensor system which was built on a 2 layer PCB having a thickness of 1.6mm and diameter of 16mm.

(a)Front-view   (b) Back-view
Figure 31: PCB of the AFE-PPG sensor system

4.3 Complete wearable measuring unit

The final wearable prototype of the measuring unit should be assembled in a way that it will be easy to use at the measurement site. The assembling task should include three major electronic components: the PPG sensor system, the IMU and the 3V battery for the power supply. Silicon conformal coating was done to the prototype to protect it from moisture, corrosion and thermal shock. Figure 32 shows the final wearable heart rate measuring unit.

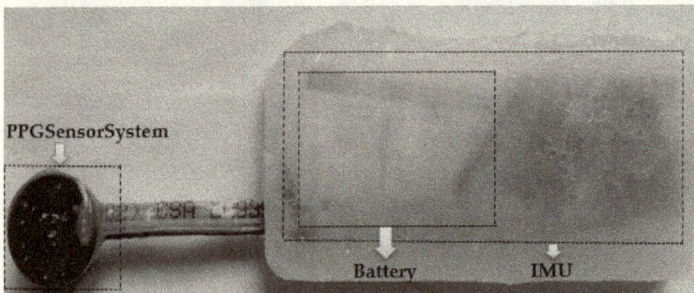

Figure 32: Complete wearable heart rate measuring unit

4.4 Software implementation

4.4.1 Sampling the AFE-PPG signal

Flowchart in Figure 33 illustrates the steps needed for sampling an analog signal by the CC430F5137 microcontroller.

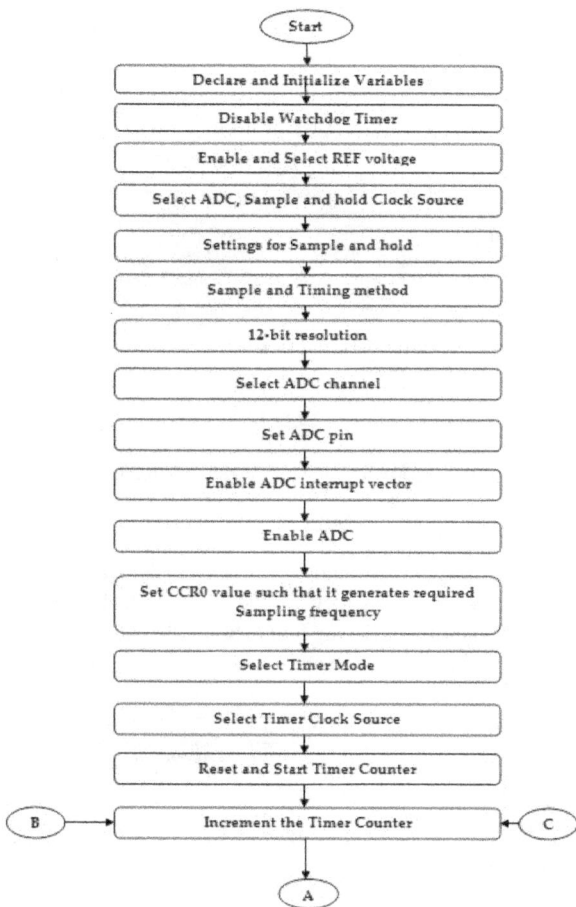

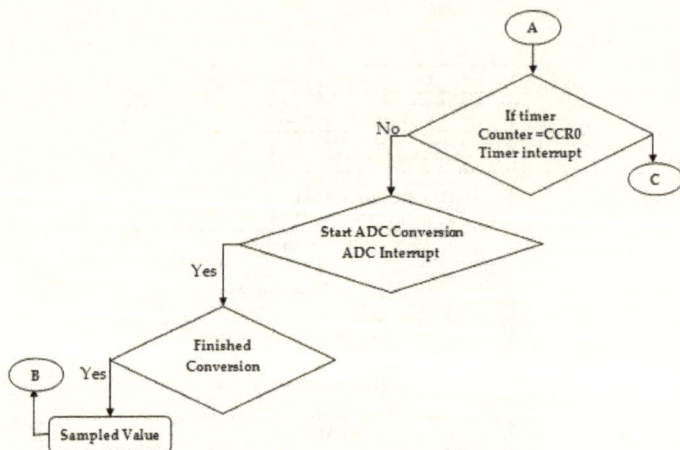

Figure 33: Method for sampling the analog signal

4.4.2 Determining frequency of the PPG signals

Of the two methods explained in section 3.5, it was analyzed that method1 is a very slow way to determine the frequency as the microcontroller detects things in a fraction of a second. The method2 was found to be the solution for this application and can be implemented as shown in Figure 34.

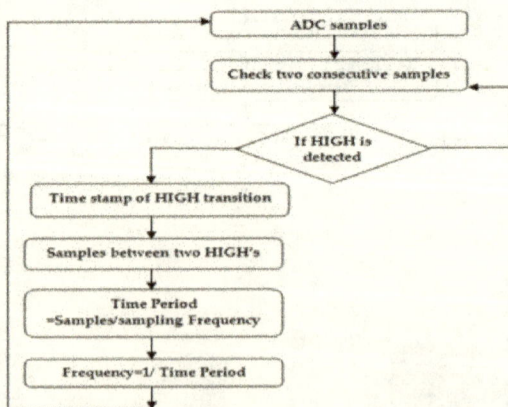

Figure 34: Determining frequency of the signal

4.4.2.1 Minimizing motion artifacts

A number of measurements were made to analyze the effect of motion on the measured PPG signal, from which it was concluded that the PPG signals were added with the frequencies lying outside the range of 0.7Hz to 2.34Hz. In order to limit the measurable heart rate to 42BPM to 140BPM the added frequencies as a result of motion artifacts are to minimized which can be done in the following ways.

1. **Removing PPG signals of frequencies greater than 3.4Hz**

The frequency components greater than 2.34Hz in the measured PPG signals will correspond to a beat-to-beat time interval of less than 0.42 seconds. Figure 35 illustrates the method to remove the frequencies greater than 2.34Hz. From Figure 36 (a); it can be observed that the detected PPG signal encounters a beat to beat time interval of 0.375 seconds (4.12sec-3.75sec) that corresponds to a frequency of 2.6Hz which is greater than 2.4Hz.

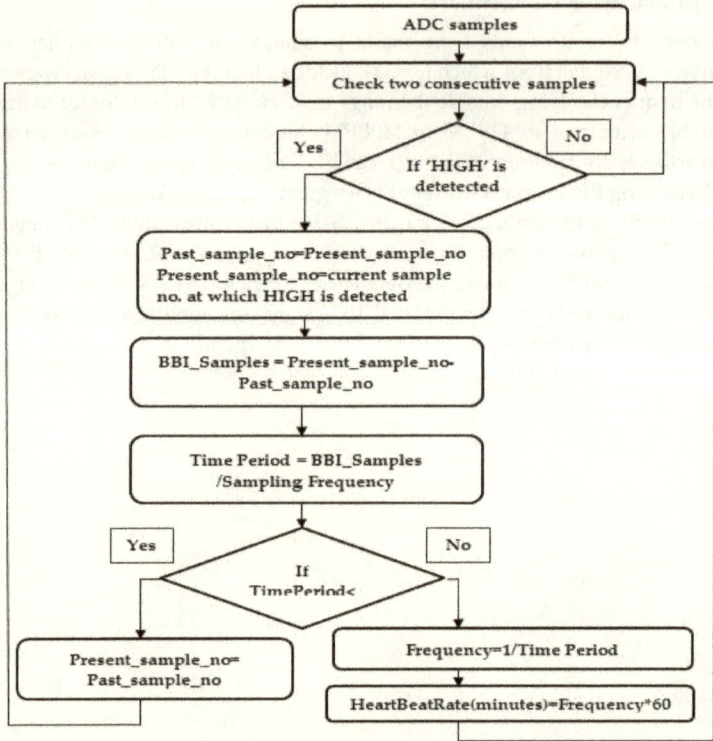

Figure 35: Method to remove the frequency components of PPG signal greater than 2.34Hz

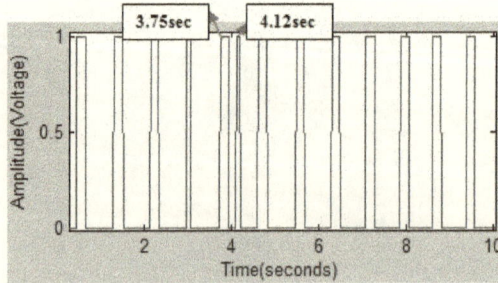

(a)Detected PPG signal

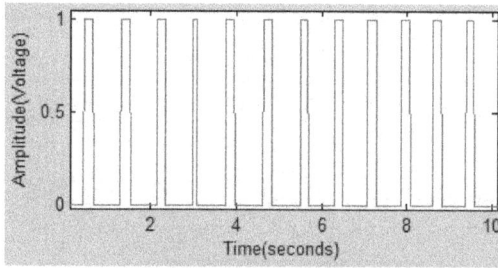

(b)Filtered PPG signal

Figure 36: Result after removing the frequency components greater than 2.34Hz

2. Removing PPG signals of frequencies less than 0.7Hz

PPG signals of frequencies less than 0.7Hz will correspond to a time period greater than 1.42secs which can be minimized by the method shown it Figure 37.

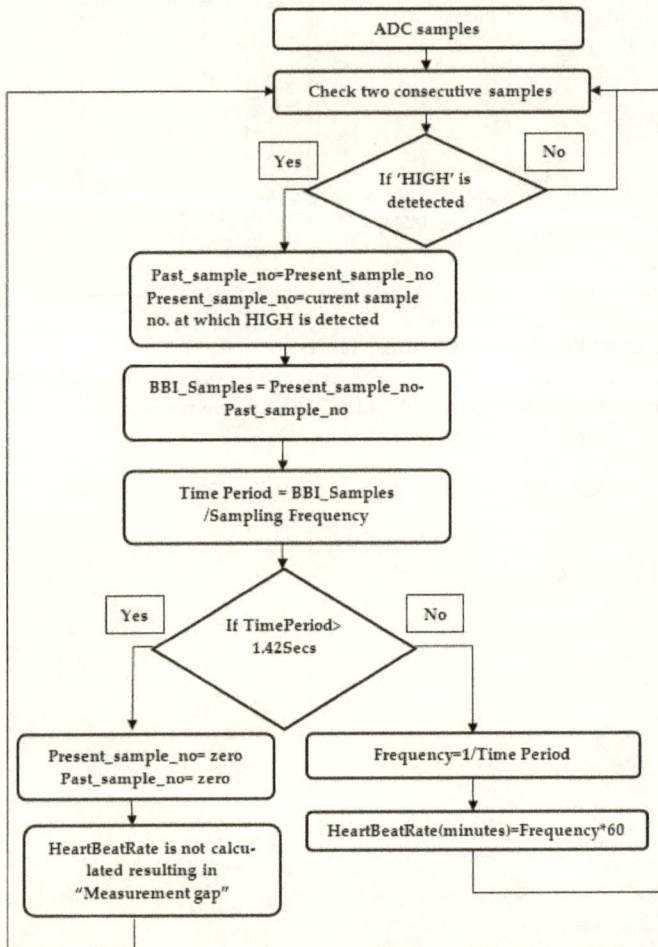

Figure 37: Method to remove the frequency components of the PPG signal less than 0.7Hz

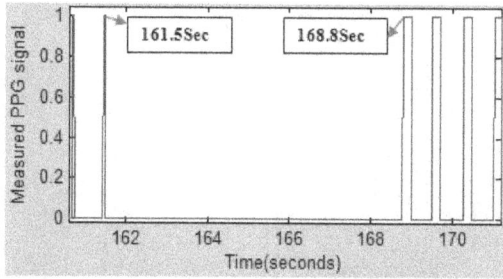
(a)Detected PPG signal

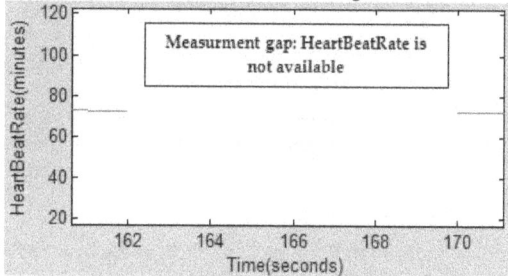
(b)Filtered PPG signal

Figure 38: Result after removing the frequency components less than 0.7Hz

From Figure 38, it can be observed that the detected PPG signal encounters a beat to beat time interval of 7.3 seconds (168.8sec-161.5sec) that corresponds to a frequency of 0.13Hz which is less than 0.7Hz. Eliminating the frequency components less than 0.7Hz that corresponds to a time period of 1.42 seconds will result measurement gaps in the calculated heart rate and eventually turning the measurement system into repair mode, which was explained in section 3.8.

4.5 Algorithm for removing motion artifacts

Methods for removing the motion artifacts as discussed in section 4.4 will be implemented both in MATLAB and on microcontroller. The working of the implemented motion artifacts algorithm on microcontroller is verified with the M ATLAB implementation to ensure that it perfectly works with the microcontroller implementation. Figure 39 shows the method for comparing both the implemented algorithms.

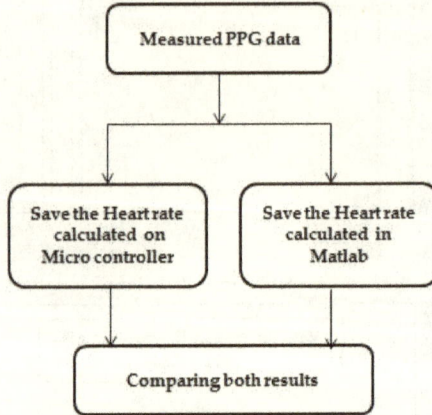

Figure 39: Verifying the implemented motion artifacts algorithm

4.6 Monitoring the measured heart rate

The measured heart rate can be monitored in the following two ways.

4.6.1 Real-time monitoring

MATLAB GUI showed in Figure 41provides a user interface for monitoring the real-time heart rate if the measurements of user are carried out by connecting the IMU to a PC via USB serial communication [Figure40]. In this method the real-time heart rate is calculated on the microcontroller from the measured PPG data. MATLAB GUI in Figure 41 shows the two plots, where the upper plot shows the measured PPG signals and the lower plot shows the calculated heart rate.

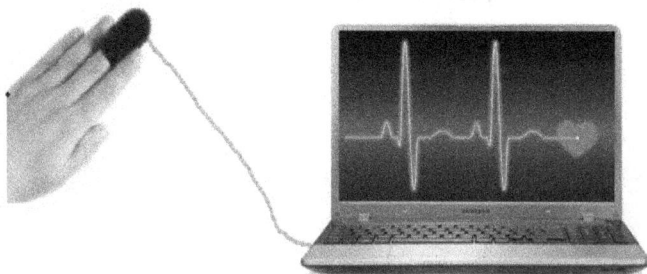

Figure 40: Monitoring the real-time heart rate using USB serial communication

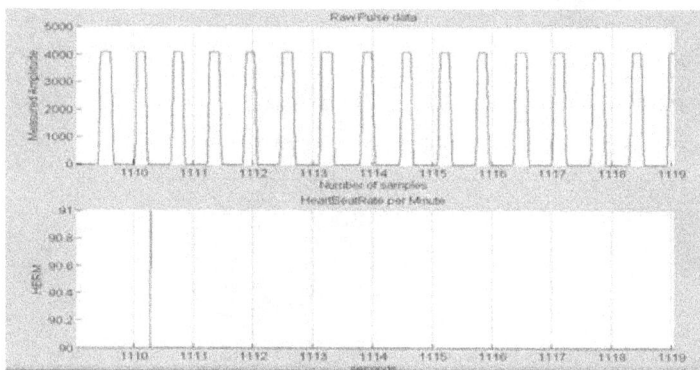

Figure 41: MATLAB GUI for monitoring the real-time heart rate and PPG signals

4.6.2 Post user activity monitoring

The developed measured unit will store the measured PPG signal data on the on-board flash memory of IMU, which can be read back into a PC by using a USB serial communication between the IMU and a PC. The data read from flash memory into PC can be post processed in MATLAB for monitoring the measured heart rate during the carried out user activity.

4.7 Validating the measurement unit performance

The following presents the efficiency of the developed measurement unit in measuring beat-to-beat pulsation when compared with the other standard devices.

4.7.1 Benchmarking

The PPG sensor working with the calculation of beat-to-beat pulsation was benchmarked with heartbeat rate measurements from other standard devices listed below.

1) The AFE4490SPO2EVM [Figure 42] from Texas Instrument (TI), measuring oxygen saturation parameters and heart rate [26].

2) Optical pulse probe and shimmer3 from Shimmer [Figure 42] measuring heart rate [27].

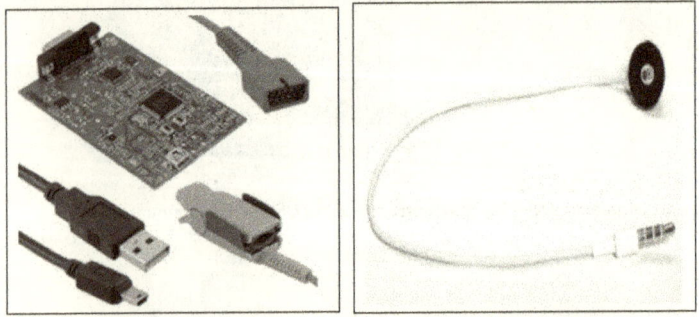

(a)AFE4490SP02EVM from TI   (b) Optical pulse probe from shimmer

Figure 42: Heart rate measuring devices used for benchmarking [26], [27]

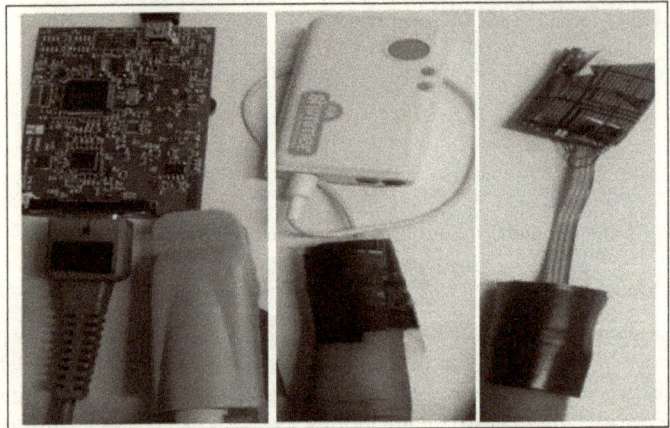

(a)AFE4400-Texas Instrument   (b) Shimmer   (c) Acreo sensor

Figure 43: Physical environment of the connected devices

Two devices were benchmarked at the same time. For benchmarking Shimmer sensor and Acreo sensor; the Acreo sensor was attached to the tip of the middle finger and the shimmer sensor was attached to the tip of the forefinger of the right hand as shown in   Figure 43.

For benchmarking Acreo sensor and AFE4490, the Acreo sensor was attached to the tip of the middle finger and the AFE4490 was attached to the tip of the middle finger of the left hand as shown in Figure 43.

4.7.1.1 Heart-rate monitoring tests

The beat-to-beat rate calculated from the measured PPG signals using the method explained in section 4.2.2 along with the heartbeat rate calculated with the other two devices with different users will be discussed here. The heartbeat rate was directly calculated from the measured PPG signals using only the method in section 4.2.2 and additionally no other special software techniques were used for smoothing or reshaping the measured PPG signals. The experiments were conducted for 60 seconds with all the three devices with their respective heart rate waveforms shown in Figure 44.

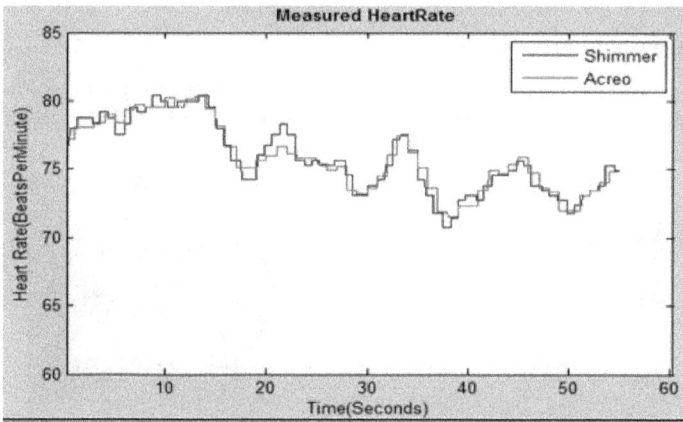

(a)Heart beat measured with finger by Acreo sensor and Shimmer sensor

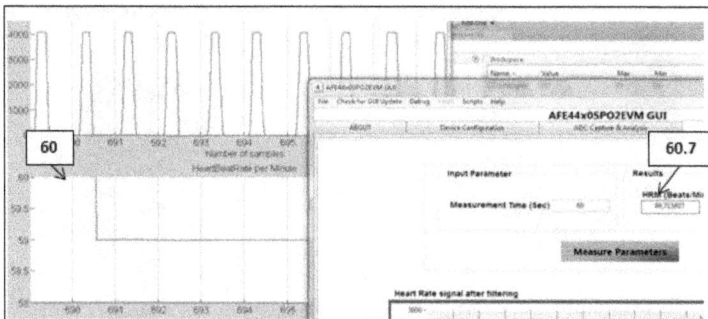

(b)Heart rate measured with finger by Acreo sensor and AFE4490

Figure 44: Comparing the heart rate measured with finger by Acreo sensor, Shimmer sensor and AFE4490

Figure 44 (a) and (b) shows the heart rate measured at different instants of time.

From Figure 44 it can be observed that the variation of heart rate measured with Acreo sensor has close correlation with that of the Shimmer sensor and AFE4490.The calculated value of Root Mean Square Error (RMSE) of the beat-to-beat heart rate of Acreo sensor and Shimmer sensor was of 1.39beats/minute, while that with the Acreo sensor and AFE4490 was found to be 0.9beats/min.

4.8 Multiple applications

Even though the prototype is being designed specifically for the swimmer application, it could be used for the other applications by changing its way of deployment on the measurement site. The following sections present the multiple applications of the sensor on different measurement sites.

4.8.1 Wrist and finger

The finger and fingertips are the traditional measurement sites for measuring heart rate in general and medical applications [15]. Figure 45 shows the physical attachment of sensor for finger and fingertip measurements.

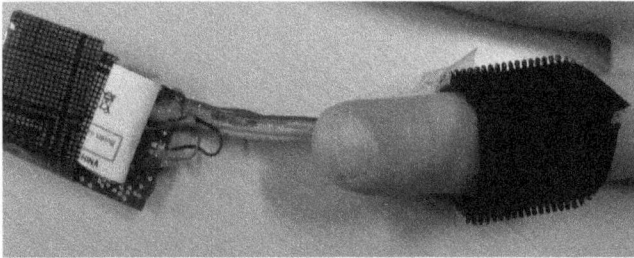

(a) Sensor attachment method for finger

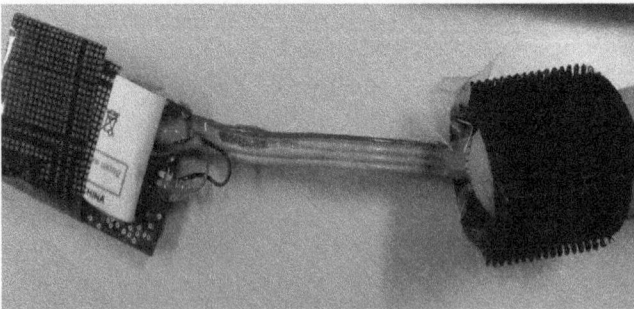

(a) Sensor attachment method for fingertip

Figure 45: Sensor attachment methods of finger

It was observed that the heart rate measurements from both finger and fingertip were correlating with the heart rate measurements of Shimmer sensor and AFE4490.

Even though both the measurement sites provide good results, each of them covers a significant portion of the finger. Thus, they cannot be applied to the users working with or using their hands.

The implementation of wrist based sensor shows challenge with the sensor placement because the arterial locations and depths at the wrist differ with users [16], [7].

The Acreo sensor developed has been attempted for the measurement at wrist as shown in Figure 45 (a) with their results presented in Figure 45 (b).

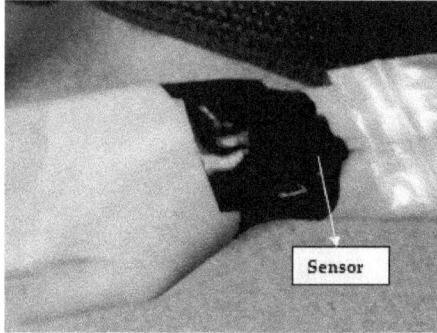

(a) Sensor attachment method on wrist

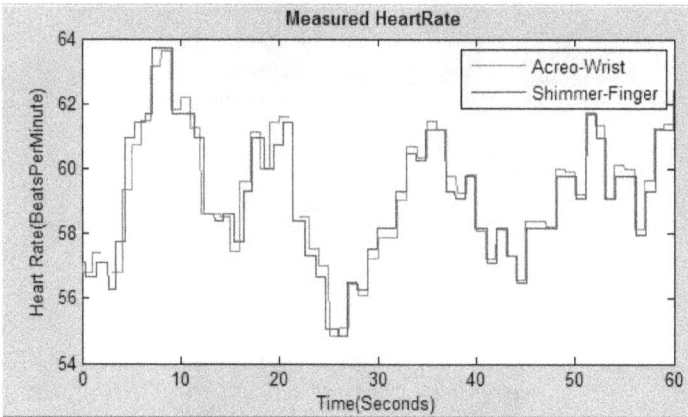

(a) Measured heart rate(Beats Per Minute)

Figure 46: Heart rate measured with finger by Shimmer sensor and

From Figure 46, it can be observed that the measured heart rate by Acreo sensor on wrist has correlation with heart rate measured on finger by Shimmer sensor.

4.8.2 Forehead

There were numerous studies involved with the investigation of the facial region for the pulse measurements, including the forehead, jaw and chin. From studies [3] and [17], it was concluded that the measurements from the jaw and chin locations are more prone to motion artifacts compared to the forehead measurements. The studies suggest that the forehead is the best measurement site as it is less prone to motion and has a sufficient density of vascular elements to provide a reading. In addition to this, its large bone structure is well suited for capturing reflected light. Moreover, the forehead measurements don't affect the dexterity of the user as they do with the finger or fingertips measurements.

4.8.2.1 Integrating sensor with a headband

The developed sensor was integrated into the existing headband as shown in the Figure47.

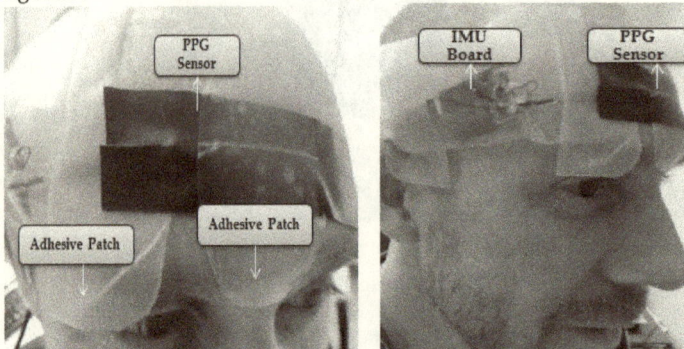

(a)Sensor attachment on forehead   (b) Side view of the attached sensor

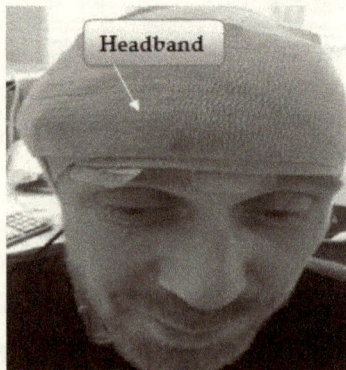

(a) Wrapping the attached sensor with a headband
Figure 47: Integrating sensor on forehead with a headband

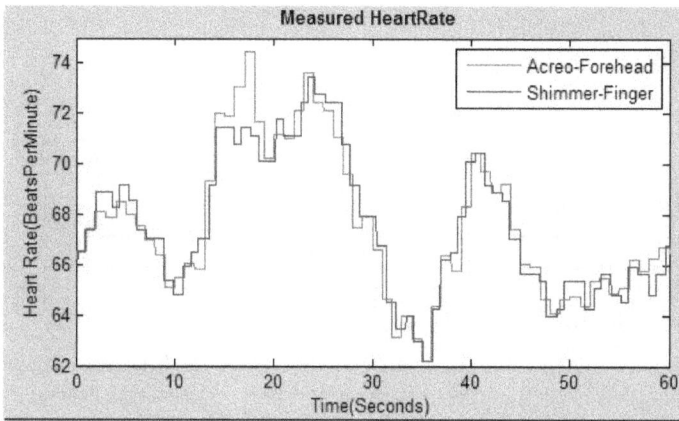

Figure 48: Heart rate measured on forehead by Acreo sensor and with finger by Shimmer sensor

From Figure48, it can be observed that the measured heart rate by Acreo sensor on forehead has correlation with heart rate measured on finger by Shimmer sensor.

4.9 Validating prototype reliability

The prototype reliability was evaluated by the user study that includes the measurements from forehead of four different users of different age groups performing different activities to understand the effect of motion artifacts on the measurements. The measured PPG data during each activity was stored on the on-board flash memory of IMU and post processed in MATLAB to analyze the measurement gaps.

4.9.1 Selection of activities and Prototype users

All the users for this study were members from Acreo Swedish ICT.

1. Sitting

This forms a basic study to understand the placement and minimal required pressure to get good readings and which forms the baseline for the later activities. This was conducted for all users to know the requirements for the sensor attachments.

2. Walking

After placing the sensor at the right place, the user was instructed to walk for 1 minute inside the Acreo office. This was conducted in order to check the effect of simple movements of walking on the measured readings; if the motion is badly affecting the reading then the sensor cannot be implemented for more rigorous activities later.

3. Jogging

This activity includes a bit more movements compared with the earlier because of the movement of the whole body including the forehead.

4. Running

Finally, running activity was selected.

Table4.1 shows the total time duration, average, total, and maximum gap times in seconds for each activity of the user.

Table 1: User activities

User	Sitting	Walking	Jogging	Running
U1	60 - - -	60 1 1 3	60 1.5 2 3	120 1.1 2 7
U2	60 1 1 1	60 3 7 9	60 2.6 5 8	60 1.5 4 9
U3	60 1 1 1	60 1 1 1	66 5.5 12 22	108 3 3 3
U4	60 - - -	60 1 1 1	60 1.6 3 5	60 1.7 3 7

4.9.2 Factors affecting the measurement system

From the user data in Table4.1, it can be observed that the measurement results will be affected either by the user or with the user activity. The factor affecting the measurement system should be determined so that it will be minimized to get a valid result.

From Table4.1, it can be observed that the user2 and user3 showed a measurement gap of 1 second while sitting which was resulted from their activity of adjusting the sensor. User3 has a measurement gap of 22 seconds that was resulted from combining the activity of stairs with jogging.

From the Table4.1, it can also be observed that the maximum measurement gap time increases for each user as we move along the table from the activity of sitting to running except with the jogging activity of user3.From this it can be concluded that the activity of the user plays a major role in affecting the measurement system. Apart from this, it was analyzed that the following factors will also result in the measurement errors.

1. Contact pressure of the sensor

It was analyzed that the contact pressure of the sensor on the measurement site plays an important role for valid readings. If the pressure is too low, then there are chances that the ambient light will corrupt the measurement data and on the other hand, if the sensor is attached by applying more pressure against the measurement site might lead to cutoff of blood flow eventually leading to the invalid PPG signals [2].

2. Measurement site

From the different experiments conducted by the different measurement site as discussed in section 4.6 it was analyzed that measuring at the wrist is very different compared to the other measurement sites because of the bone structure at the wrist that doesn't hold the sensor properly.

4.10 Monitoring the heart rate of swimmer

Total wearable measurement unit prototype is attached on forehead by following Figure 47, which was then followed with integrating the swimming cap as shown in Figure 49.

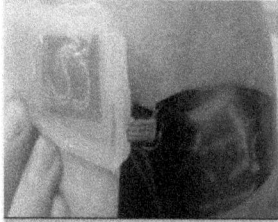

(a)Attaching the measurement   (b) Integrating with swimming cap

unit on forehead

Figure 49: Integrating entire measurement unit on forehead with swimmer cap

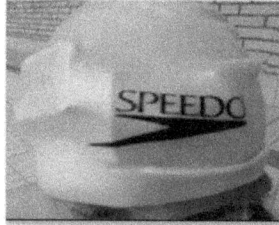

4.10.1   Testing the entire prototype during swimming

For monitoring the heart rate of swimmer the method 2 "Interval sets with a specific amount of time or rest between swims" suits the best from the two methods explained earlier in section2.9.1, therefore the swimmer was instructed to follow it. The swimming activity was carried out with 4x50's freestyle (meters) with leaving some time interval of rest between each 50-meter as shown in Figure 50, and the time took for each action is tabulated in Table 4.2.

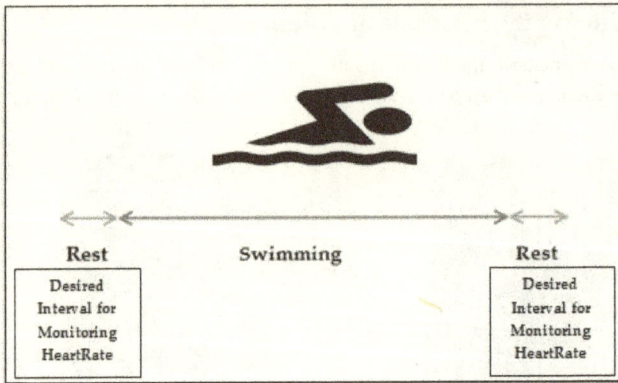

Figure 50: Desired interval for monitoring heart rate of the swimmer

4.10.1.1 Measured PPG data

During the swimming session the measured PPG data was stored on the flash memory on the IMU and post processed in MATLAB for removing the motion artifacts and calculating the heart rate.

The measured PPG data during the entire swimming session is shown in the upper section of Figure 51 with Table4.2 showing the time interval for each activity, duration of the activity (seconds), time interval (seconds) and their respective color representation in MATLAB plot.

Table 2: User activities and their time duration

Activity & Interval	Time Duration (Seconds)	Time Interval (Seconds)	Representing Color in the Plot
Resting +Walking + getting into the pool	208	0-208	Black
Resting interval1	20	208-228	Blue
Swimming interval1	45	228-273	Red
Resting interval2	27	273-300	Blue
Swimming interval2	48	300-348	Red
Resting interval3	30	348-378	Blue

Swimming interval3	52	378-430	Red
Resting interval4	40	430-470	Blue
Swimming interval4	46	470-516	Red
Resting interval5	39	516-555	Blue
Getting out of the Pool+ Walking+ Removing the Sensor	119	555-674	Red

4.10.1.2 Calculated heart rate

The measured PPG signal data for the entire swimming session is shown in the upper part of Figure 51, while the lower part of it shows the calculated heart rate.

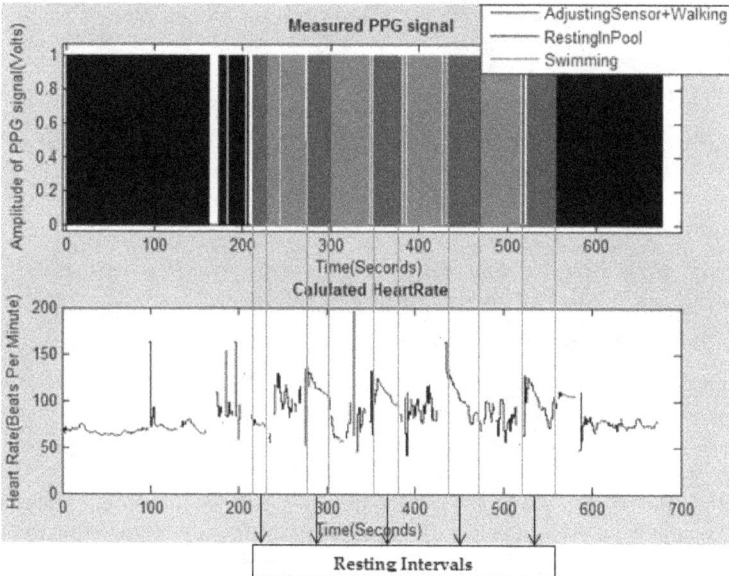

Figure 51: Heart rate measured during the entire swimming session

The following presents the analysis on the variation of the measured heart rate for each activity of the entire swimming session.

1. Resting and walking into the pool

The data for this activity interval is presented with the black color plot in Figure 51, which is shown again separately in Figure 52 with the measured heart rate during this activity. From Figure 52, it can be observed that the heart rate measured is valid measurements until 163 with some measurements gaps in between and from 163 seconds the measured heart rate suddenly increases or decreases which was resulted from the corrupted PPG data. The measured PPG data was corrupted because of the associated action of swimmer while getting into the pool.

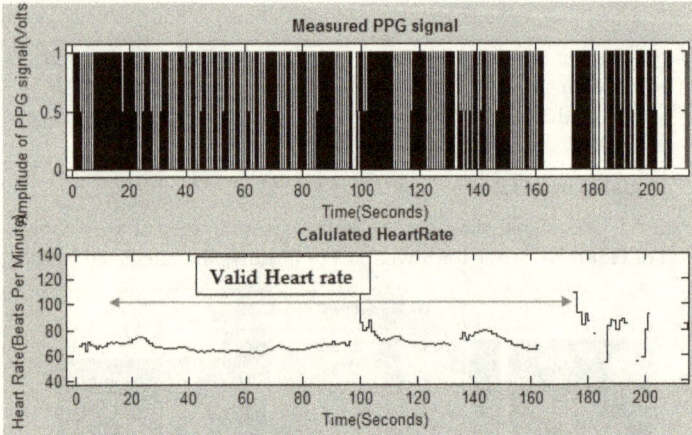

Figure 52: Measured PPG signals and heart rate during the walking

2. Resting

From Figure 51, it can be observed that during every interval of resting, the measured heart rate follows two phases as explained below.

a) Sudden changes in the heart rate

This phase1 represents the sudden changes in heart rate from the start of every resting interval. The heart rate measured during this phase is not a valid measurement because the swimming activity is associated with the motion artifacts that results in the corrupted PPG signals. During the change of activity from swimming to resting, the sensor needs some settling time to overcome the motion artifacts and to produce a valid measurement. This settling time may not be the same for every resting interval as explained below with different resting intervals.

> **Resting interval1:** Phase1 of this interval has a settling time of 6 seconds.

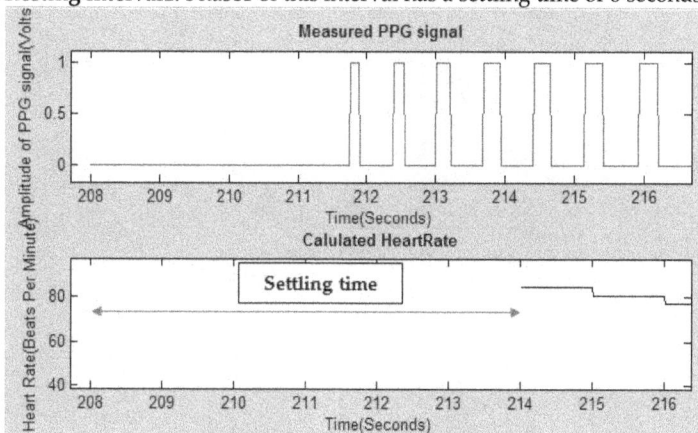

Figure 53: Phase1 of resting interval1

> **Resting interval2:** Phase1 of this interval has a settling time of 2 seconds.

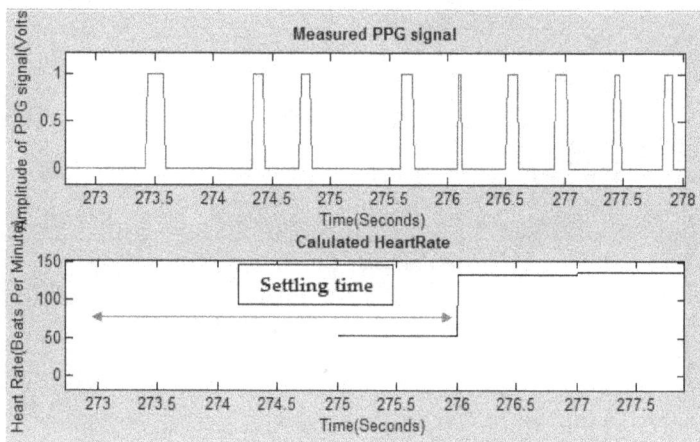

Figure 54: Phase1 of resting interval2

> **Resting interval3:** Phase1 of this interval has a settling time of 9 seconds.

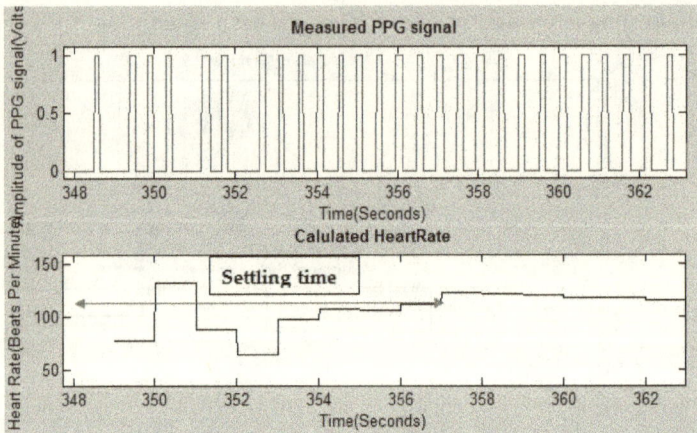

Figure 55: Phase1 of resting interval3

> **Resting interval4:** Phase1 of this interval has a settling time of 6 seconds.

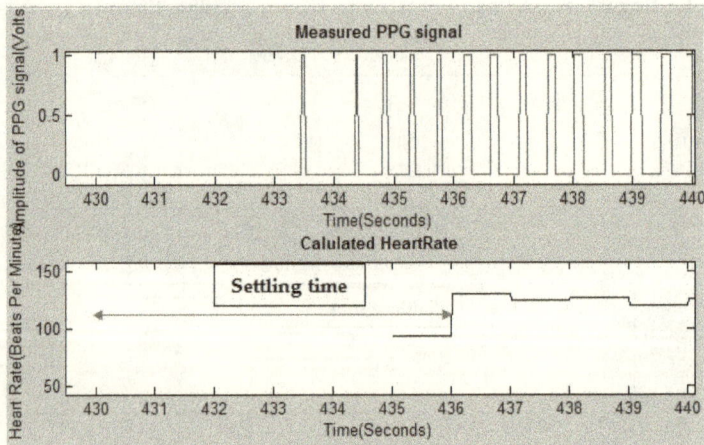

Figure 56: Phase1 of resting interval4

> **Resting interval5:** Phase1 of this interval has a settling time of 8 seconds.

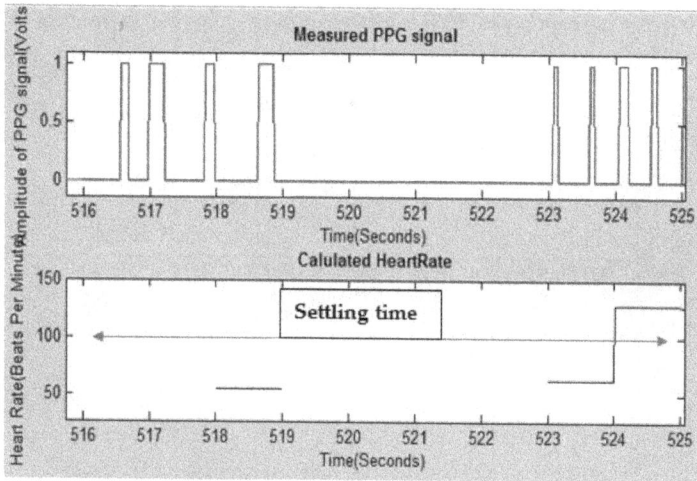

Figure 57: Phase1 of resting interval5

b) Gradual decrease of heart rate

This phase2 represents the gradual decrease of heart rate followed after the phase1.The heart rate measured during this phase is a valid measurement because of the fact that the heart rate increases during swimming and then decreases during resting.

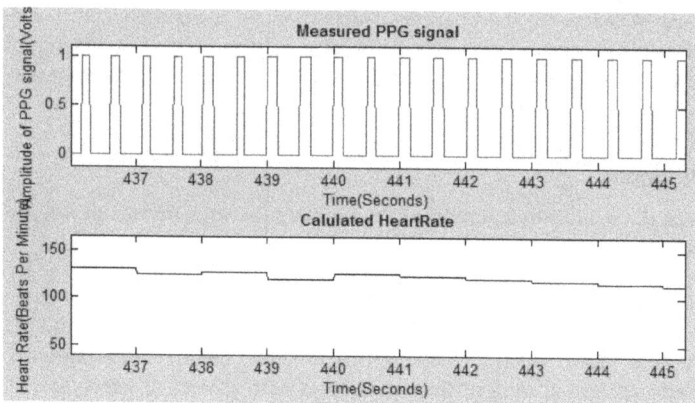

Figure 58: Phase2 of resting interval4

3. Swimming

From Figure 51, it can be observed that the heart rate measured during every swimming interval follows a zig-zag path and is not a valid measurement, the reason for this being the corruption of PPG signals resulted from the swimming action as shown in Figure 59 with the swimming interval3.

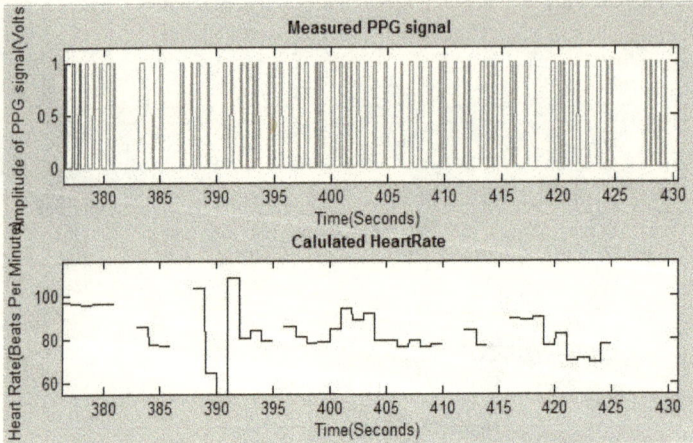

Figure 59: Measured PPG signals and heart rate during swimming interval3

From Figure 59, it is clear that the measured PPG signal is corrupted with the motion artifacts resulting in the measurement gaps and invalid heart rate readings.Conclusions and Future Work

Many challenges existed throughout the project and many workaround techniques were used to overcome the problems. After months of research with the development of on this measurement unit, the final conclusion is to answer the following questions which can be considered as the conclusions and future work for this project.

4.11 Outcome

What was the overall outcome from the month's research on this prototype development?

The prototype of wearable heart rate measuring unit was developed and during resting it was benchmarked with AFE4490 and Shimmer sensor at multiple measurement sites, from which it was observed that the heart rate measured with the developed unit was having a close correlation with that of AFE4490 and Shimmer sensor. From the carried out user study for the activities like running, jogging, walking it was observed that the unit was exhibiting measurement gaps of only seconds in the measured heart rate that doesn't seriously affect the measurements on a large-scale.

For measuring the heart rate during swimming the measurement unit was integrated into the swimmer cap, from which it was observed that the unit was measuring heart rate during the resting intervals of the entire swimming session.

From above, it can be concluded that the developed prototype forms a solution for measuring heart rate at different measurement sites and also measurements with both resting and moving activities.

4.12 Future work

What kind of future work will be associated with the carried out work?

As the developed measurement unit is a prototype, its future work can include the following

- **Improvements with the measurement unit prototype design**
-

    The developed prototype of measurement unit needs improvements with the design like miniaturization and better packaging as shown in Figure 60 such that it is both desirable to the end-user while providing a quality data. Also the design needs a solution for maintaining an optimal pressure against the measurement site.

Figure 60: A smart-patch prototype for embedding the measurement unit

- **Mobile application for real-time monitoring of the vital signs**

    In general, vital signs refer to the measurements of basic functions of human body. The main vital signs basically monitored by medical professionals and health care providers are :

    1) Body temperature

    2) Heart rate

    3) Respiration rate (rate of breathing)

4) Blood pressure (Even though it is not considered a vital sign, it is often measured along with the other vital signs)

The emerging smartphone technology led to the increase of app-based mobile heart rate monitors. There exists many HRM's that provide a mobile application for monitoring the vital signs using wireless communications like radio frequency or Bluetooth. But some of them provide the mobile applications for monitoring the vital signs during some specific activities like swimming, running, cycling, etc. and the other HRM's available for monitoring the vital signs during swimming exhibit some issue as explained in section ,therefore they cannot be used for monitoring the vital signs during swimming. Hence, there is a real need to developed a mobile application for real-time monitoring of the vital signs and also other parameters like the start and stop time for each swimming lap, time taken for each lap.

- **Improving the measured heart rate during motion artifacts**

The problem of measurement gaps associated with the heart rate of the developed measurement unit can be overcome by collecting the PPG data simultaneously from two sensors placed at different measurement sites [29].

5 References

[1] J.A.C Patterson, "A Photoplethysmography System Optimized for Pervasive Cardiac Monitoring," Imperial College London, February 2013.

[2] R.Dresher, "Wearable Forehead Pulse Oximetry: Minimization of Motion and Pressure Artifacts," Worcester Polytechnic Institute, Massachusetts, United States.

[3] A.Nagre and Y.Mendelson, "Effects of Motion Artifacts on Pulse Oximeter Readings from Different Facial Regions," *Proc. of the IEEE 31st Annual Northeast Bioengineering Conference*, pp. 220-222, 2005.

[4] P.D.Mannheimer, et al., "The Influence of Larger Subcutaneous Blood Vessels on Pulse Oximetry," *Journal of Clinical Monitoring*, 18, pp. 179-88, 2004.

[5] A.C.M.Dassel, et al., "Reflectance Pulse Oximetry at the Forehead Improves by Pressure on the Probe," *Journal of Clinical Monitoring*, 11(4), pp. 237-44, 1995.

[6] R.P.Dresher andY.Mendelson,"Reflectance Forehead Pulse Oximetry: Effects of Contact Pressure During Walking"

[7] Q.Cai, J.Sun, L.Xia, and X.Zhao, "Implementation of a Wireless Pulse Oximeter Based on Wrist Band Sensor,"Collage of Biological Science and Medical Engineering, Southeast University, Nanjing, China.

[8] J.Spigulis,M.Ozols,R.Erts, and K.Priditis,"A portable device for optical assessment of the cardiovascular condition," University of Latvia,Physics Department and IAPS, Latvia. Available: http://home.lu.lv/~spigulis/PPG-AOMD-3.htm

[9] J.B.Forsyth,T.L.Martin,D.YCorbett, and E.Dorsa," Feasibility of intelligent monitoring of construction workers for Carbon Monoxide Poisoning,"Virginia Polytechnic Institute and State University, Balcksbur ,VA.

[10] NJRC products for Health care. Available: http://semicon.njr.co.jp/eng/PDF/NJL5303R_E.pdf

[11] B. Roberts,"Capturing Grid Power,"*IEEE Power and Energy Magazine*, vol. 7, pp. 32-41, 2009.

[12] Colin," Converting an analogue signal to a digital signal".

Available: http://users.tpg.com.au/users/talking/a_to_d.html

[13]   R.Bhatt, "A DIY Photoplethysmographic sensor for measuring heart rate".
         Available: http://embedded-lab.com/blog/?p=5508

[14]   J. B.Forsyth,"Wearable Pulse Oximtery in Construction Environments," Virginia Polytechnic Institute and State University, Virginia, March29, 2010.

[15]   Nonin Medical Inc.,"Nonin fingertip oximeters,"2010. [Online].

         Available: http://www.nonin.com/Index.aspx

[16]   K.Li," Wireless Reflectance Pulse Oximeter Design and Photoplethysmographic Signal Processing," B.S.Zhejiang University, 2010.

[17]   W.Johnston, P.Branche,C.Pujary, and Y.Mendelson, "Effects of motion artifacts on helmet-mounted pulse oximeter sensors, " Proceedings of the IEEE 30th Annual Northeast Bioengineering Conference,April 2004.

[18]   S.Rhee, B-H.Yang, and HH.Asada, "Artifact-Resistant Power-Efficient Design of finger-Ring Plethysmographic Sensors,"IEEE Transactions on Biomedical Engineering, Vol.4, No.7, July 2001.

[19]   Polar,"Heart Rate Monitor and GPS watches".   Available:
         http://www.polar.com/en/products

[20]   H.Kent ,"Understanding Interval-based Training in the Swimming Pool".
         Available: http://www.trinewbies.com/tno_swim/tno_swimarticle_06.asp

[21]   Suunto,"Suunto Accessories of Heart Rate monitors".
         Available: http://www.suunto.com/Product-search/Accessories/

[22]   Suunto,"User guide for Suunto Memory belt".
         Available:
         http://ns.suunto.com/Manuals/Memory_Belt/Userguides/MemoryBelt_UG_E N.pdf

[23]   M.Butcher ,"Instabeat Is Revolutionary HUD For Swimming Goggles". Available:http://techcrunch.com/2013/05/16/instabeat-is-revolutionary-hud-for-swimming-goggles-you-can-back-on-indiegogo/

[24]   Instabeat-Hud for Swimmers.
         Available: http://www.instabeat.me/blog/2014/04/08/instabeat-hud-swimmers/

[25]   Sunnto,"Suunto ANT Heart Rate Belt".
         Available: http://www.suunto.com/Products/sports-watches/Suunto-Ambit2-S/Suunto-Ambit2-S-Red-HR/?categoryId=3

[26]   Texas Instruments," AFE4490 Evaluation Module for Pulse Oximeter Applications".
       Available: http://www.ti.com/tool/afe4490spo2evm

[27]   Shimmer,"Heart Rate measurement using Optical Pulse Probe ".
       Available: http://www.shimmersensing.com/shop/shimmer-optical-pulse-probe

[28]   Texas Instruments," CC430 Family-User's Guide".
       Available: http://www.ti.com/lit/ug/slau259e/slau259e.pdf

[29]   J.ee and J.Nam," Design of Filter to Reject Motion Artifacts of PPG Signal by Using Two Photo sensors," J. Inf. Commun. Converg. Eng. 10(1): 91-95, Mar. 2012.

[30]   J.Perez,"Aplications and implications,"FAB ACADEMY 2013.
       Available:
       http://academy.cba.mit.edu/2013/students/contonente.javier/week16/week16.html

*All URL links are available on 13th of May 2014.